Anthropogenic Rivers

EXPERTISE

CULTURES AND TECHNOLOGIES OF KNOWLEDGE

EDITED BY DOMINIC BOYER

A list of titles in this series is available at www.cornellpress.cornell.edu.

Anthropogenic Rivers

The Production of Uncertainty in Lao Hydropower

Jerome Whitington

Cornell University Press
Ithaca and London

First published 2018 by Cornell University Press

Printed in the United States of America

Library of Congress Cataloging-in-Publication Data
Names: Whitington, Jerome, 1974– author.
Title: Anthropogenic rivers : the production of uncertainty in Lao hydropower / Jerome Whitington.
Description: Ithaca : Cornell University Press, 2018. | Series: Expertise : cultures and technologies of knowledge | Includes bibliographical references and index.
Identifiers: LCCN 2018027133 (print) | LCCN 2018035347 (ebook) | ISBN 9781501730924 (pdf) | ISBN 9781501730931 (epub/mobi) | ISBN 9781501730900 | ISBN 9781501730900 (cloth ; alk. paper) | ISBN 9781501730917 (pbk. ; alk. paper)
Subjects: LCSH: Ethnology—Laos. | Human ecology—Laos. | Sustainable development—Social aspects—Laos. | Hydroelectric power plants—Environmental aspects—Laos. | Hydroelectric power plants—Social aspects—Laos. | Dams—Environmental aspects—Laos. | Dams—Social aspects—Laos.
Classification: LCC GN635.L28 (ebook) | LCC GN635.L28 W45 2018 (print) | DDC 306.09594—dc23
LC record available at https://lccn.loc.gov/2018027133

The book is dedicated to my parents, who introduced me to Southeast Asia, taught me to see the living environment, and showed me what an enigma people can be.

Contents

Preface

Anthropogenic Rivers: The Production of Uncertainty in Lao Hydropower is premised on the observation that ecology has become central to the status of anthropos in the contemporary.[1] While this *problematique* has been a long time in coming, it is now clear that there is a decisive, transnational interest in examining the significance of human-induced ecological change within popular culture, political movements, governments, and even business. Essential to this emergence has been a shift from the comparatively discourse-centric risk politics of the 1980s and 1990s to what we can call an ontological politics of uncertainty. Risk politics hinged on an epistemological dilemma and was preoccupied with how to make socially valid decisions when safety could not be guaranteed. It also appeared strikingly Eurocentric because it maintained belief in rational decision making and state guarantees of safety, while retreating into the sense of security offered by experts. Even in the United States, risk politics frequently took the form of perception management (the Tylenol poisonings), pop culture (apocalypse; Don DeLillo's novel *White Noise*), or outright environmental injustice (chemical alley; Hurricane Katrina)—not the formal problem of a technocratic rationalism that could no longer trust in the ideology of value-neutral truth claims. For marginalized communities around the world, the expectation that policy decisions would guarantee their safety seemed like a bizarre premise to begin with.

By contrast, the ontological politics of uncertainty is characterized by powerful actors who strategically produce uncertainty (for instance, by undermining scientific truth claims); by deep-seated disinvestment in knowledge infrastructures; by uncertainty as "built in" to ecological relations themselves (rather than as only a discursive or epistemic problem); and by the destabilized temporalities of anthropogenic natures, infrastructures, and knowledges (in which people are obliged to speculate on dangerous futures or attend to the latent effects of the industrial past). Among other things, uncertainty is the domain of opportunistic commercial actors who protect themselves against threats while taking advantage of risky possibilities—oftentimes producing yet more instability in the process. Uncertainty is the result of a capitalist valorization of long-odds achievement that does not care very much about theoretical justifications for action and is premised on the logic that demonstrated achievement, not correct representations of reality, is the only real proof of the worth of an idea. With deep resonances with American pragmatist philosophy, it is a radically different orientation than that of a rationalism that hides value judgments behind authoritative scientific claims. The contention of this book is that late industrial environments are constituted by an ontology of uncertainty in which actual ecological relations become deeply uncertain, and that this condition is essential for debates about the status of the human vis-à-vis global ecologies.

By exploring this extensive "production of uncertainty," I show that the anthropogenic relation should be construed as a double movement. Typically, the term "anthropogenic" refers to human-induced ecological change and raises questions about culpability and the distribution of harm within human and nonhuman ecological relations. But much contemporary practice in fact explores the creative potential of destabilized ecologies and asks how new capacities for being are emergent within pervasive ecological change. This view, which is different from the positivist, determinist anthropology of Jared Diamond–style reflection on our ecological predicament, implicitly posits that the human is not defined as a moral entity or biological species, but rather as a work in progress in a condition of permanent emergence within tense and fraught ecological relations. If we understand "the human" not as a transcendental subject or as a species defined by universal characteristics, but as an emergent work in progress, then the anthropological question shifts from "what does it mean to be human?" toward something more like "what are people capable of?." What distinctive human formations are possible now or in the future? This question is necessarily speculative, for it is only as a matter of inventive engagement that it is possible to venture an answer. "The endeavor," writes Anand Pandian (n.d.) is "to conceive a humanity yet to come." Moreover, it insists that the capacity to inflict and bear harm should be included within anthropologists' theorization of the human, just as much as, say, cutting edge science (e.g., Rose 2007). Hence, unlike debates about the Anthropocene, a metanarrative that appears inevitably foreclosed and committed to an unworkable notion of civilization (see esp. Scranton 2015; Oreskes 2014), I venture to suggest that radical ecological change posits the necessity of open-ended experimentation on the human, fraught with the risk of failure, and that such experimentation at the level of practice is in fact what is taking place. Anthropos is the dependent variable in an unknown techno-ecological function. Witness troubling geoengineering experiments currently underway or Elon Musk's commitment to climate change entrepreneurialism and Mars colonization. Those are not the only kinds of experiments possible, but the fact remains that we Earthlings, human and otherwise, simply do not know who we are going to become.

To address these ideas, why turn to a relatively obscure sustainability experiment in the borderlands of Southeast Asia? The immediate context of the research was an experimental collaboration between a major hydropower company operating in Laos—probably the most profitable company in the country at the time—and a well-known transnational activist group, International Rivers. What made this collaboration distinctive was that it was an overwrought political situation and yet, on the other hand, it was also a genuine experimental collaboration in which people were taking real risks and attempting to concoct new kinds of formations. This was true also of the villagers, farmers, and fisher-people who were coming to terms with major livelihood transformation and the experimental sustainability designs of the private-sector hydropower company. Moreover, at that time (in the mid-2000s), Laos was itself the subject of experimentation on large-scale hydropower design and financing. In short, this was a situation in which diverse groups of people were working to accommodate anthropogenic rivers into their practices and dispositions in novel ways. If Lao rivers were becoming newly anthropogenic, so too were people and their diverse institutions "becoming riparian" in the sense that their experimental practices sought to accommodate the river in new and committed ways. Many of the current debates surrounding ecology address the global, and they do so in terms that evoke a similar "global" generality. By contrast, it is important to remember that ecological subjectivation frequently takes place at the level of comparatively mundane practices, and it often has little to do with ecological conscious-raising or ethical ideals about protecting the earth. The fact that there is so much of this mundane subjectivation, taking such extreme, plural and specific forms, is the challenge before us—not the creation of a global metadiscourse. If the objective is "to think what we are doing" (Arendt 1998, 5) or, more expansively, to explore a "historical ontology of ourselves" (Foucault 1997c, 315), then perhaps it is worthwhile to look at a case in which the people involved were themselves experimenting, challenging their own political commitments, and thinking aloud about what they were doing and why. The result, therefore, is an attempt to develop tools and problems (Rabinow 2003, 2008), rather than a theory or narrative of all-encompassing change.

Finally, the part of my approach that I think will prove most jarring to academic readers is the attempt to think about the contingencies of late industrial environments through the ontological terms posed by capitalist actors themselves. "Uncertainty"—as a domain of practice, as a way of looking at the world, and as an affirmation of open-ended contingency—is essential to the materialist ontology of managers, technicians, and experts, and this book can be taken as an attempt to "ontologize up" rather than to take for granted the "Great Divide" (Latour 1987) that is frequently invoked to debunk claims of authoritative certainty. Martin Holbraad, Morten Pedersen, and Eduardo Viveiros de Castro might put it thus, when they write that the challenge is "to pass through what we study . . . releasing shapes and forces that offer access to what might be called the dark side of things" (2014). I offer an experiment in postnatural grounded theory much in the way that the term "postcolonial" refers not to the end of colonialism but to the pervasive inheritances of colonialism after its formal end. To this extent, I am less interested in a theoretical critique of the concept of nature and more interested in the experimental elaboration of postnatural ecologies. Capitalists, after all, are materio-semiologists par excellence, and the Foucauldians and actor network theorists have been silenced, or at least forced to retreat into the safety of knowledge institutions, in the face of current political events. I only ask, in a very classical anthropological gesture, that the reader inhabit this position for a while in order to understand what it feels like.[2] Postnatural ecologies look quite different from this vantage—more open-ended, less concerned with authoritative knowledge that we might presume, superficially more optimistic, and also quite a bit darker. At a moment when we are called on to vouch for the authority of environmental sciences and institutions in the face of a full-scale capitalist assault on environmental knowledge, I hope this ethnography will serve as a detailed look into one capitalist modality of power/knowledge and its ontological implications.

A note on sources and anonymity: all ethnographic informants in this book are treated anonymously, with two exceptions: Ian Baird and Bruce Shoemaker, two scholar activists who played a role in the events I describe, agreed to be identified in the text. Village names were changed, but the names of companies and NGOs are left intact. All quotations should be considered paraphrased from longhand notes.

Acknowledgments

Working in Laos was a challenge and I am grateful for the many acts of kindness and support that made it possible. Thanks to the Theun-Hinboun Power Company for permission to conduct research; CARE International in Lao People's Democratic Republic; International Rivers Network; Paul Cunnington, Geraldine Zwack, Charles Alton, Charles Jenneret, Houmphanh Rattanavong, Grant Evans, Ian Baird, and Bruce Shoemaker; the engineers at Sogreah, Nam Theun Power Company, Electricité du Lao, and the Ministry of Industry and Handicrafts; and the team at Resource Management and Research. Thanks to the many affected villagers for trusting me enough to share their stories, and to the Environmental Management Division technicians for answering obtrusive questions and letting me observe their work. I was welcomed with undue hospitality and assisted by many people who unfortunately cannot be named here. I only hope I have done justice to their stories.

For their intellectual mentorship and unflagging support I would like to thank Aihwa Ong, Lawrence Cohen, Nancy Peluso, and Paul Rabinow at the University of California, Berkeley. Many thanks are also due to Cori Hayden, Donald Moore, Peter Zinoman, and Ashley Thompson, also at Berkeley. At the University of Texas, Katie Stewart, Ward Keeler, and others will detect notes of their abiding influence. I would also like to thank Meg Stalcup, Alfred Montoya, Tobias Rees, Janelle Lamoreaux, Joshua Craze, and Jelani Mahiri. For reading and commenting on drafts of the manuscript, I thank Alberto Sanchez-Allred, Daromir Rudnyckyj, Joe Bryan, Ben Gardner, Jerry Zee, Casper Jensen, Andrew Johnson, and several anonymous reviewers—the book is immeasurably better for their insights.

Thanks to Trevor Paglen, Tania Li, and Anand Pandian for much-needed encouragement to begin writing; Michelle Murphy, Kim Fortun, Stephen Collier, and Michael Fischer for sage advice; and, at Cornell University Press, editor Jim Lance and series editor Dominic Boyer for believing in the project and for astute guidance.

Research for this book was funded by the Fulbright-Hays DDRA program and a University of California Pacific Rim research grant. I learned a tremendous amount during a fellowship at the Climate Justice Research Project at Dartmouth College, and I can only express my pleasure at getting to work with Michael Dorsey, Gerardo Gambirazzio, and Lauren Gifford. Writing was made possible by the Asia Research Institute of the National University of Singapore, and deep appreciation is due to Greg Clancey, Prasenjit Duara, and Jonathan Rigg for their generous support and mentorship. Thanks to many friends and supportive colleagues in Singapore. At New York University, I am very appreciative of the generosity of the Department of Anthropology, as well as earlier support from Eric Klinenberg and the Institute for Public Knowledge. Parts of my argument were presented at Stockholm University, Johns Hopkins University, Kyoto University, McGill University, and Columbia University; many thanks to the numerous participants for insightful comments and questions.

It has been an immense pleasure to write this book at the same time that my partner was writing hers, and I can think of no greater satisfaction than to do so while raising our two daughters. The gratitude owed to one's parents is immeasurable, and I can only hint here at the deep love, thanks, and joy that comes with getting to share this book with my mother and father, to whom it is dedicated.

Anthropogenic Rivers

Introduction

The Production of Uncertainty

I have in my possession an audio recording that captures very well some key dimensions of my field research in the Lao hydropower industry in the mid-2000s. For much of the research I worked closely with a consulting team of environmental scientists in their work on the environmental planning and impact assessment for large dams in Laos. The team was run by an aged British man and several of his sons; they made the recording, and I discovered it in a small archive of field research materials and project documents. The recording itself says a lot about Lao hydropower, but here I am only concerned with the relation between the environmental consultant and two managers of the hydropower company, Richard, a manager, and Buali, head of its Environmental Management Division.

The setting is the hydropower company's field office, at the base of a mountain ridge at the edge of a small highway in the central part of the country. I spent months there on stints from Vientiane, the capital city, during 2003–2005, trying to understand the forms of reason surrounding what some people wanted to call sustainable hydropower. The meeting among the three of them was held in one of the office's bare concrete rooms. In the recording, the underpowered air conditioners wheeze and groan trying to keep up with the humidity. What captures for me the predicament of Lao hydropower development is the difficulty with which the British consultant fails to extract some basic facts about the company's environmental management program:

Consultant:	I mean that is *the* important question. I'm just saying that the most important question here is—are the people better off than they were before?
Buali:	No, no. . . . We can't say ourselves you know. Oh, if you ask me now I can tell you [it's] very good—they are better off, of course. Because [they have] more rice to eat.
Consultant:	It's your intuition.
Buali:	Yeah, that's what I think. If you measure it.
Consultant:	I mean, I would like to see the socioeconomic baseline data. And has there been—there's no report on socioeconomic indicators?
Buali:	No no.
Consultant:	When you're doing your work—
Buali:	We collect the GDP data year by year. In some villages [unintelligible] the first year—
Consultant:	Only some villages.
Buali:	Yeah we're not able to [go back more than] 3 years.[1]

I listen to passages again and again, trying to make out certain words. It is an allegory for the fieldwork as a whole: facts that never materialize, data only sporadically collected, speech that becomes unintelligible at its most important, the opacity of relationships (why these recordings?) and meetings I was not allowed to attend. One lesson of the recording is that there is no back room meeting that explains the secrets of the Lao hydropower industry. There is no recording that documents the truth and clarity of its power relations. There are not even claims of fact—only gestures and opinions. I feel like the fashion photographer in Michelangelo Antonioni's film *Blow Up*,[2] who discovers in the background of one of his photographs an apparent image of a murder—but who must enlarge the photo again and again, trying to discern the image more clearly, until the blurry murder surrenders to the grain of the medium.

The dialogue in the recording hinges on the consultant's demand for a basic understanding of the dam's effects on villagers' livelihoods. What was their original living status? How much has their income changed due to the effects of the dam and five years of trying to improve their condition? At what point will the company's responsibilities be fulfilled? In short, what in fact is the reality of the dam for villagers?

Trending

Richard was a hydropower company manager and a mediator par excellence. In the recording, Richard offers a consummate performance of his skill at mediating delicate relations by repeatedly claiming that things are going great, that they have really had a lot of success in fixing the dam's problems and villagers' lives are rapidly improving—all the while qualifying his praise in muddled undertones, persistent lack of context, spin after narrative spin most reliably summed up by the noisy connotations of the word *trending* (his emphasis):

Consultant:	You've got a lot of baseline data—I'm just wondering—you've only got three years in some villages, two in some others, but you also have some [pre-construction] baseline data, very sketchy—
Richard:	Some but not much. Not enough to do a comparison from what I've been told.
Consultant:	Ok, so you just have those sets of data—from what I've been told there's no analysis that's been done yet. Because you don't have enough information.
Richard:	We're—we're—we're—we're *trending* it to see if there's anything. And on some indicators it looks like yes, [unintelligible]. On malnutrition under five, that indicator seems to have been a huge improvement in villages we've started work programs in.
Consultant:	Ok, that sounds great!
Richard:	And so . . . Dr. Nouanchan, she—she's [unintelligible]. You can't just use one [indicator], but that's a decent indicator to show whether an average family is doing better.

Consultant: Sure.

Richard: And somewhere [unintelligible] villages 5.9%. . . . So we can say, although [unintelligible], well, we started at this point three years ago, now what we've done is we've gotten them here.

At every moment when Richard is unintelligible in this recording, he purports to hedge his claims of good news; to qualify a claim for which there is no data; to provide evidence, context, or background information; or to excuse an indicator that is not really relevant to the discussion. The consultant needs to know something basic: are the villagers any better off after attempts to fix the problems with the dam? Yet image games are the central management strategy—even with the consulting team working for his company on a contract directly relevant to a matter-of-fact representation of the situation. In sum, Lao hydropower is "after nature," in two respects: the reason of hydropower planning is no longer organized with respect to knowledge of the objective world as the substantive basis for authoritative action, and yet it is still *in pursuit* of knowledge, attempting to find its ground or foundation, while identifying flexible relations that are capable of manipulation (Strathern 1992). There are no facts on the ground. The spin, the multiple asides, and the unfounded assertions are confusing, and confusion is his game. Everything is [unintelligible]. There is only trending.

Richard was hired by the hydropower company to deal with transnational activists, especially one US group in particular, International Rivers Network (IRN), which had staged a campaign against the company in the late 1990s. The Theun-Hinboun hydropower project, which is the subject of this ethnography, is owned and operated by the Theun-Hinboun Power Company, which is a joint venture of the Lao government with Scandinavian and Thai investors. The dam was completed in 1998 with major backing by the Asian Development Bank, but by 2000, IRN had successfully forced the company to acknowledge its environmental problems, commit to a major increase in environmental funding, and commission a ten-year environmental management plan. Conducted from 2003 to 2005, my research was targeted at this turn of events.

This book, therefore, addresses a broader question concerning sustainable hydropower as a renewed articulation of environmental protection and economic growth. Laos during the late 1990s and early 2000s was the target of much transnational activity concerning whether large dams could successfully navigate the shoals of new forms of highly strategic environmental activism. The predicament of Richard's company captured well the material politics in which the industry overall found itself, and his embodied, performative labor constituted in part a diagnosis of that situation and a demonstrative attempt to rekindle the material seductions of hydropower. Many in the broader expatriate development community in Vientiane, and people in the government to a lesser extent, believed that Richard and his team were keen to address the dam's problems systematically and directly. Others were suspicious of greenwashing, and the very claim of sustainability raised questions about "what was really going on." To prefigure the analysis, I found that the company's environmental interventions were extremely interesting and definitely worth paying close attention to. They did not rely on an authoritative form of scientific environmental knowledge—at least not straightforwardly—but on a kind of flexible managerialism that was distinctly American, experimental, and open-ended. Yet far from simply fixing environmental problems in a technocratic mode, the company's operations relied on and even actively sowed a certain amount of chaos.

Most provocatively, Richard the mediator had controversially initiated a collaborative arrangement with IRN, which had been a central activist force over almost two decades of global antidam organizing. My research was able to observe this collaboration between a major hydropower facility and, at the time, perhaps the most aggressive and successful antidam nongovernmental organization (NGO) operating globally. It became clear that this collaborative move was an attempt to co-opt activists into the image repertoire of a "reformed" company. Yet to make matters more interesting, the company was *also* doing a tremendous amount of work in villages to deal with the problems it had caused, and it was impossible to write off their work as mere window-dressing.

What could be going on? Why collaborate with an aggressive activist group rather than a mild-mannered development NGO that would be careful not to offend anyone? And how did this one case, dominated by Anglophone managers, activists, and experts, speak to the multiple contexts of Lao development? What did Lao development experts, officials, and villagers think of this predicament? What sort of global assemblage did this amount to (Collier and Ong 2005)? In the early 2000s, Theun-Hinboun Power Company was the most profitable company in Laos. Was the situation typical? Such questions motivated my research.

I conducted research with environmental consultants; hydropower technicians and managers; villagers living along the affected rivers and near the powerhouse and dam sites; and urban Vientianites, development workers, and officials. While focused on this dam project, I was more broadly concerned with the meaning of development, the renewed role of hydropower, and the tensions between the national frame of development, transnational developmental demands, and rural riparian lives.

One central claim of this research is that sustainable hydropower rests not on the discursive construction of authoritative knowledge or expertise, but on the active production of uncertainty. *Uncertainty* describes a tactical relation to knowledge, a condition for action or of not knowing how to act, as well as a predicament of disenfranchisement in the material conditions of infrastructure and environment. More than just the deliberate proliferation of misleading discourses, this production of uncertainty takes place first and foremost in the ecological effects of large dams. There is a definite connection between Richard's performative management, the absence of reliable information about the effects of the dam, and the conditions of lived ecological uncertainty that increasingly characterize the environmental citizenship of Lao farmers and fishers living along these transformed rivers. The problems caused by large dams are problems that the experts do not know how to fix; still less are they comprehensively known and analyzed at the level of sustained, rigorous research.

Uncertainty is not the opposite of knowledge but a constitutive relationship that acknowledges the role and value of knowledge to projects of living. It operates affectively as an often deeply felt apprehension, in which people know enough to imagine the different ways dangerous situations may turn out. Unlike risk in the sense of social distribution of probabilities, such as with insurance, uncertainty is qualitative rather than quantitative, and it is not necessarily the object of formal expertise.[3] More than a simple future-oriented anticipation, uncertainty is experienced as an ecological predicament with biopolitical stakes, as an understanding that existing knowledge is *not good enough*, and from practices that, whether they want to be or not, are open to that predicament. Ultimately, uncertainty is a relation that indicates the value of knowledge.

In my narrative, uncertainty takes form as *threat* and *opportunity*, promises, fears, and aspirations. Consider "greenwashing," those environmental claims that play on expert knowledge by presenting deliberate misrepresentations that are technically true but deeply disingenuous. There is an industry and a history around this explicitly *interested* mode of knowledge production, often backed by business interests such as mining (Kirsch 2014) or fossil fuels (Oreskes and Conway 2010). Greenwashing is fascinating not because corporations dissimulate but because, as an opportunistic, technically sophisticated practice, it forever plays with the limits of *what one can get away with* or, put differently, *what specific discursive relationships might make possible*. Promises often work similarly; the uncertainty entailed in a promise is an integral part of its seduction.[4] What might people expect from the resettlement promises of new homes and farmland? Activist claims also often raise more questions than they answer when their reliability may be suspect yet they cannot be easily dismissed. A promise or a threat may be assessed in terms of its potential and the experience it engenders—its capacity to affect the subject—rather than whether it is right, wrong, or simply an incoherent distraction. In a postfact world, even if the subject is unsure what to accept as true, the tendency is to operate "as if" it were true if it poses a threat or suits one's purposes. Anxieties and aspirations dominate. The representational tether to reality is loosed, for one is forced to imagine.

For managers, attention to threats and opportunities does not require any particular distinction between political, commercial, and technical activity, nor does it rely on sophisticated forms of probabilistic reasoning. Instead it is personalistic, intuitive, and informed by decades of accumulated managerial reason. These kinds of speculative limit practices, subject to failure and oriented toward threats and opportunities, are unlike the naturalizing performances of the construction of facts or scientific planning. Rational planning performs foundations and guarantees. This kind of entrepreneurial familiarity with threats and opportunities performs possibilities and risky achievements.[5]

"Foundational" and "achievement"-oriented knowledge practices are distinct in form but can clearly overlap in practice. A second claim of the research is that sustainability politics—the activity of activists and managers—can be understood with reference to a practice of *technical entrepreneurship* oriented toward specific material-semiotic relations. Technical entrepreneurialism is any technically sophisticated practice designed to exploit or manifest the uncertain potentiality of specific, real relations. Thus activists exploit weaknesses in hydropower planning institutions to achieve opportunistic political goals. Environmental technicians, as I show at length, both view an entrepreneurial attitude to be important for affected villagers to deal with the problems they face and consider themselves to be motivated, entrepreneurial actors who do not attempt to reform villagers' social practice but rather use those practices against themselves, as it were, to achieve developmental results. As a concept, technical entrepreneurialism refers to technically sophisticated practice that is comfortable working through risk and indeed takes opportunity and threat as its condition of possibility.

Foundational knowledge practices assume that a shared understanding of reality can ground political negotiation, whereas achievement-oriented practices do not hold this assumption and take for granted that explicit political spin within a destabilized empirical context must constantly form a key dimension of activity. Hence they operate across different political ontologies, one that engages in terms of correct or incorrect representations of reality, the other in terms of direct work on real relationships, including discursive relations. There is frequently a habitual tendency toward narrative spin, which in turn does not ground a truth politics but rather creates conditions for a certain kind of manipulative play. The manager's skill comes not from knowing objectively the environmental conditions in which his company operates, but rather from knowing how to anticipate and manipulate the diverse sociotechnical relations in which the company is enmeshed. Not necessarily negative, "manipulation" functions as a symbol for how to work on relations of all sorts.

Yet uncertainty is not simply a feature of the experts and managers' understanding. Just as human designs are built into natural landscapes, so too the uncertainties, ignorance, and inadequacy of infrastructural development become part of their ecological legacies. As the authoritative control over hydropower development has become less and less tenable, certain actors have found it important to forego attempts to control expertise or insist on naturalized facts in favor of a position that is comfortable with and even exacerbates the uncertainties faced in building dams. In doing so, they affirm and enact a central feature of anthropogenic rivers themselves: dammed rivers are ecologically novel and constantly in flux. They are new entities that bear a distinctive anthropogenic mark and that in turn force people to change themselves. It is essential, ultimately, to view hydropower experts and managers as people who, at varying degrees of remove, have extensively made themselves into riparian creatures.

My argument is not that behind greenwashed claims of environmental sustainability we will find gross misrepresentation of the facts.[6] Rather, we often will not find many facts at all because no one has invested in the knowledge infrastructure and labor necessary for making facts.[7] We find instead a deliberate predicament without adequate evidence and without ways to adjudicate knowledge claims. Sustainable hydropower development engages a form of technical practice that relies on, accentuates, and actively engages in the production of uncertainty. Qualitatively distinct from the authoritative knowledge of conventional development planning, such technically sophisticated practice is an essential aspect of capitalist enterprise. Hence, technical entrepreneurialism and the production of uncertainty have become crucial to privatized, neoliberal development. For these reasons, the grounded theory that arises out of explaining empirical problems for a limited project in Laos has broader relevance for understanding the anthropogenic potential of late industrial capitalism.

Much of what counts as sustainability in Lao hydropower and beyond falls within an unstable, uncertain field of nature-intensive capitalist development. "Trending" consists of glossy brochures and optimistic images as well as water pumps for gardens, village credit schemes, and pig pens—each of which seeks to exploit the possibilities of underdetermined relations. It consists of antagonistic contestation no less than attempts at collaboration, quiet complicities, and carefully negotiated compromises. It consists in muttering the bad news and promoting the good news. Life in these entrepreneurial zones articulates around people who constantly rethink conditions of late industrial environments and their ongoing, enterprising work on the specific relationships that define its problems. Its interventions lead to investment in results-oriented, risky reconfiguration of vital, material relationships. The language of "techno-fixes" or "techno-optimism" is insufficient. These material practices are not antipolitical because they mask or dissimulate a truer political relationship. They are *actually* political insofar as they build out a vital, formative reality that would not otherwise exist, while actively undermining the intelligibility of other worlds.

Institution Hacking and the Limits of Expertise

Anthropogenic Rivers is an ethnography of the limits of expertise, what happens when people do not know what to do, or when they know little about environmental relations that matter. It is an ethnography of the changing obligations to a river that emerge when things do not turn out as planned. It is also an ethnography of work—in the sense that ill-defined, unstable environmental relations become targets for continued work by enterprising, entrepreneurial actors. Things talk (Daston 2004). But what they signify is often their excess, and sustainable hydropower entails hope for progress and the imagination of disaster. Contestation frequently marks points of disagreement about how to understand reality, while it also displays work on relations through which unsettled, multiple realities take form and are enacted (see Barry 2001; Mol 2002). Yet one could also say that reality itself becomes deformed in time, as when speculation emerged that the devastating 2008 earthquake in Sichuan, China, may have been triggered by the massive volume of water filling the recently finished Zipingpu Dam. The speculative possibility is that "added weight both eased the squeeze on the fault, weakening it, and increased the stress tending to rupture the fault" (Kerr and Stone 2009, 322). Contestation is only one social form among many possible ways in which people work through uncertain political and ecological implications of major infrastructural projects. Nonetheless, the strategic contestation of transnational activism has played a crucial role in the changing politics of knowledge in sustainable hydropower.

The Theun-Hinboun Dam project faced a situation that emerged in the late 1980s to characterize many large, internationally funded hydropower projects. In fact, Laos was taken to be a model site for the emergence of a new approach to hydropower development, in part because of the degree of rural political control and the absence of organized civil society, in strong contrast to India, Brazil, or Thailand where pitched battles over large dams were waged in the 1980s and 1990s. The dams themselves were also different by design, leading to smaller inundation zones and, with Laos's low population density, far fewer people to be resettled. By the mid-1990s, it was increasingly difficult to fund or build large dams on conventional financing. In neighboring Thailand, due to rural popular mobilization and the rise of a counterdiscourse of local ecological knowledge, it had become impossible. Sustainability became an export market overnight as plans for

Lao dams gained added favor, and the main customer was the Electricity Generating Authority of Thailand, Thailand's public utility. Laos's promises of sustainable hydropower were articulated against the limits of development expertise and the infrastructure investment that expert confidence once underwrote.

Late in 2004, I watched a development expert throw up his hands in dismay and frustration during a public consultation workshop as an activist asked an acerbic question about his work. The expert had spent the previous eight months redesigning his company's elaborate social and environmental plans for a large dam it was preparing. It was perhaps a month before I departed the capital of Laos, rife with its development politics and the urban tenor of a kind of hip environmentalism, when I walked into the broad foyer of Vientiane's diplomatic and corporate landmark, the Lao Plaza Hotel, to attend the public consultation workshop. The workshops were mandated by the World Bank as part of a requirement to solicit public opinion. The hydropower developers had learned they needed to spend a lot of time talking to people not directly involved in their dam project if they ever wanted it to get off the ground.

The public consultation was to present to Vientiane's development practitioners and government technocrats a host of plans meant to mitigate the environmental impacts of the World Bank's flagship of sustainable hydropower, the Nam Theun 2 Dam in central Laos (see Goldman 2005). The dam had become the site of continued negotiations between transnational environmentalists and development organizations. Somehow the camps seemed to cleave to each other around an implicit question. What should be done in those villages affected by the dam to mitigate its environmental impacts? After affixing my name tag, I was handed a sort of briefcase embossed with the name of the workshop and the company logo. In it were multiple packs of printed PowerPoint slides, a public relations portfolio, and a CD containing a number of expert reports and videos of villagers participating in planning meetings. The struggle for green hydropower development had forced a mimetic profusion of planning and an exponential proliferation of its artifacts—the stuff of sustainability.

The frustration of the experts during this meeting was no less predictable than the distrustful air of the activist. "What do you expect us to do?" one could imagine the expert saying, exasperated. The sense of a failure of knowledge was aptly captured by one observer in an Internet posting concerning the Xayabury Dam project in Laos, who expressed deep mistrust of project developers and the knowledge produced:

> How people trust you if you failed to follow or ignore the rules set forth? This mega project is not only for individual or a single group of people, but it is for all Lao citizen as well as indigenous people in the neighboring countries who will, one day, lost their agriculture activities, culture and livelihood after flooding by the dam. Who know? Can EIA [environmental impact assessment] or SIA [social impact assessment] tell? Can the results of those EIA and SIA be disseminated for the public? How reliable [are] the EIA and SIA results conducted by the project developers?[8]

Mistrust in knowledge and in social relations here go hand in hand. As Paul Rabinow writes, "the social authority of experts has been undermined by their oft-proven inability either to forecast the future or to make it happen as envisioned" (2008, 59). The dam promoters had spent the previous four or five years ensuring that the project design encompassed the detailed environmental criticisms of transnational NGOs. The sheer number of documents, reports, studies, and action plans was testament to the destabilized role of expertise. The stacks of plans and kilograms of studies—as the experts themselves emphasized—still had to be supplemented with an elaborate public relations performance, and indeed it was not a performance of the authority of expertise. The layers of thick images of sustainable hydropower were less the confident longings of an updated midcentury developmentalism than a crafted argument reflexively aware of its audience's skepticism.

The developers' performance was a response, on activists' terms, to their insistent questioning of the limits of technology in relation to livelihoods struggles. Political controversy and critique have become incorporated within the technocratic and financial practices through which dams are built and operated. Stakeholder participation is a managerial form (Fortun 2001); it recognizes that the modern god trick of a singular, objective view of the world is a liability because it promises too much responsibility over nature. Stakeholder participation substitutes dialogic murkiness for univocal assent increasingly difficult to maintain in practice. Better to allow for controversy, managed within limits. Controversy is a method or social form through which the limits of expertise are internalized in knowledge practices. In the planning process, the experts were obliged to concern themselves with the perspectives of marginalized villagers somehow ironically recentered as tokens of public debate. The specific valence of that irony was harder to determine.

It may in fact be tragedy, not irony, at stake in the pluralization of perspective. The social role of activism is to verbalize the limits of expertise when it comes to technological effects. Questions of *voice* become apparent when experts acknowledge their challenges, failures, and limits. Try to hear the tenor of Richard's speech act: "We're—we're—we're—we're *trending* it to see if there's anything. And on some indicators it looks like yes, [unintelligible]." The gesture is a performative hedge. A hedge is a risk management position. Similar to an excuse, a hedge is "how to not exactly do things with words."[9] Many of the problems that confront large hydropower dams are problems that cannot be solved. The performative hedge indexes both the threat of a political situation and the poorly understood flux of not-natural ecologies.

Furthermore, contestation does not simply concern the proliferation of angry questions and voices about the limitations of expert knowledge. Environmental activism in the 1990s became very sophisticated about its material tactics for undermining knowledge infrastructures of hydropower development and planning. Their pervasive material tactics created a situation in which hydropower managers and experts were potentially vulnerable at every joint in the network of relations that made their work possible. That threat of activism is one of the key ways in which the production of uncertainty set the stage for sustainable hydropower. Environmental knowledge and development expertise could no longer be relied on in the same way.

Activists' work was distinctly material in that they sought to undermine the conditions through which hydropower development was possible, primarily by targeting specific institutional processes such as project approval and expert environmental knowledge. For example, activists' field-based research demonstrated unacknowledged problems with the dam. When the industry tried to respond to these claims, industry actors ended up revealing how little they knew about the riparian environments they had damaged—effectively proving activists' point. In another case, demands for better and more detailed plans trapped the company in a reductio ad absurdum of interminable demands for more environmental knowledge. The pervasiveness of these technically sophisticated practices set the stage for an industry that felt it was vulnerable at potentially any point along the planning networks it had long taken for granted. Whereas "environment" was previously something external to most of the company's operations, activists thrust environmental risk into potentially any number of crucial company operations. Environment ceased to be an external context to become a pervasive condition of relations themselves. In short, activists demonstrated the extent to which environmental risk saturated the institutional practices of hydropower development.

This institution hacking—a kind of technical entrepreneurialism—often consists in turning the very presuppositions of institutional expertise against it. The weaknesses and aporia of expert systems are exploited in an active, materialist form of deconstruction. Large hydropower dams create a tremendous range of social and ecological uncertainties along the rivers where they are built, but it is activists—in part channeling that very ecological uncertainty into new forms of political relation—who have systematically problematized hydropower development and denaturalized its institutional conditions of existence. A whole variety of sustainability interventions or strategies has been formulated to address this experience of vulnerability. The frustration of the experts, the pathos of expertise, and the glossy stuff of sustainability are the hallmark of an industry obliged to confront its untenable expert assumptions—not in vague, general terms, but at potentially every meeting, with each new report and each scene of intervention and negotiation. Sustainability and controversy go hand-in-hand, as it were.

Correlatively, the entrepreneurial ethos of activists and managers alike entails familiarity of working with risky, real relationships to discern what possibilities they might entail. (From the French *entreprendre*, to undertake, it need not refer to commercial activity.) Enterprise means trying something out, committing to a process, seeing what works and what does not, and remaining open to revising one's values and objectives. It is opportunistic in that opportunism is oriented toward what may or may not come to pass and is therefore taken in an affirmation of risk and experience. By recognizing at least implicitly that ecological futures cannot be planned for, controlled, or predicted with any finality—which does not mean planning or prediction are useless—entrepreneurial practice invests in specific, committed sociotechnical relations without guarantee and without knowing what surprises they might entail.

In the Theun-Hinboun power project, I document the collapse of a certain commitment to planning and the subsequent emergence of a regime of managerial reason that forms a central, entrepreneurial relation to uncertainty. Management invokes a long history of pastoral power relations but, just as importantly, involves a range of specific material activities that do not add up to an overarching, predetermined plan. Rather, a number of open-ended material and discursive practices attempt to ground a novel affirmation of hydropower. Sustainability practice, heavily invested with technical entrepreneurialism, concerns this multiplicity of experimental relations as an exploration of the possible, from new energy technologies to novel discursive formations. Motivated, enterprising subjects are essential to the postplanning engagements of sustainable hydropower because nature is conceived in terms of potential, not in terms of necessity or limit. In the apprehension of opportunity and threat, the practical question is *what do these real relationships make possible?*

Anthropogenic Rivers

Imagine a twenty-seven-meter-high dam spanning a narrow cleft in a low mountain ridge (figure 0.1).[10] Once upon a time, perhaps not long ago, a river cut through this cleft to exit a plateau of rolling moraine hills, sandy soils, and brush forest interspersed with clusters of rice fields. Early in the rainy season, one could smell the subtle ash and low charred stumps from the fields cleared annually before the rain; the occasional paddy field with low bunds to be repaired and waiting algae, which thrived as soon as the monsoons saturated and softened the earth. Rivulets, nearly dry much of the year, quickly filled with darting spawn as well as young people with nets and sometimes organized teams with bamboo fish traps. Here, small clusters of villages forged a living through rotating cultivation of hill rice, fishing and hunting, riverbank gardening, and periodic sale of forest products to itinerant traders circuiting through from Vietnam or sometimes Thailand. During the war years, American bombers flew sorties toward Hanoi over a depopulated river zone too dangerous for habitation while the river-loving communities stuck to the hills for safety. Now the dam has turned the river into a reservoir of sorts. Technically it is called a headpond since, when full, it does not go above

Figure 0.1. "The imaginary and the real figure each other in concrete fact" (Haraway 1997, 2). People living in the vicinity of dams are acutely aware of the immensity of nationalist developmental projects that dwarf their ecological and political existence. Overlooking the dam with a local village leader (*left*) and research assistant. Photo by author.

the established flood line from before construction. In the rainy season the water is visibly flowing—it is not really a lake, but of course it is not a river either. But the most profound changes have come from much more intimate engagements, and the ethnography of this drainage raises questions about the viability of life here.

Imagine a dam spanning some 250 meters across this cleft, holding back a volume of water weighing millions of tons. The design matters. Hydraulically, the Theun-Hinboun Dam is a transbasin diversion scheme—the water is channeled from a high elevation river through a five kilometer tunnel beneath a mountain ridge, for an elevation drop of some 230 meters before it exits the turbines into a much smaller river that carries it on to the Mekong. Two hundred thirty meters of elevation; 110 cubic meters per second water volume; 210 mega-watts generating capacity; 17 percent financial return—you get the picture. Number expresses difference like no other register (Guyer et al. 2010) and, to borrow a phrase from Henrietta Moore (1994), hydropower relies on a passion for difference. Nearly all of the electricity from this dam is exported to provincial cities in northeastern Thailand. The project's foreign financial backers are Thai, Norwegian, and Swedish. The Lao government owns a controlling share, while key managers are Anglophone—American, Canadian, and New Zealander. The dam performs a material axis of difference and asymmetry: before/after, above/below, gravity/electricity, and Laos/Thailand. The action occurs along these axes of potential.

The most important asymmetry in my story is the difference between the incredible technical sophistication of hydropower technology, wrought through decades of capital-intensive investment and innovation, and the paucity of rigorous investment in the "knowledge infrastructure" capable of addressing life along the rivers affected by dams. As Dominic Boyer argues, "Electricity . . . culminates a centuries-long project of science, engineering, and design to capture the earth's electrical phenomena and domesticate them through generation, conduction, and insulation, taming something like the sudden explosive power of a lightning strike or electrostatic shock into something steady, reliable, and unremarkable" (2015, 532; see also Hughes 1983). By contrast, the research and even basic science required to address environmental justice concerns remains woefully inadequate, underfunded, and simply "undone" (Frickel et al. 2010; Kleinman and Suryanarayanan 2013), resulting in the social production of ignorance when it comes to ecological concerns that call into question economic practices. Infrastructures, including knowledge infrastructure, take time to develop and depend on a long-term accretion of investment and expertise.

Infrastructure also induces a powerful temporalization that is integral to the built form of dams themselves. People living in the vicinity of dams are acutely aware of the immensity of these nationalist developmental projects that dwarf their ecological and political existence. Politicians, planners, and financiers are likewise wrapped up in and enraptured by their promises and their pitfalls, both of which index the underdetermined potential of powerful technologies. Promise and pitfall are both temporal terms that help mark how late industrial technologies open up a future in which reality itself is an open-ended question. In this vein, Susan Star and Karen Ruhleder (1996) encourage anthropologists to ask, "when is infrastructure?." The question underscores that all built technological forms have histories, are subject to decay and rot, and have potentially vast effects that must be engaged as part of their stories. It may take years, decades, or more to understand the implications of a dam.

What temporalities remain latent in a river's changing confluence of forces? Building a dam is subject to temporal conditions of impermanence, irreversibility, and unpredictability (Arendt 1998), related to what Bruno Latour (1999, 281) describes as the "surprise of the action" that accompanies engineered and scientific domains.[11] To bring together all of these diverse and singularly powerful elements into a single configuration is to lash together powers that barely stay together, whose forces pull in all directions and require continual attention: the developer, engineer, or manager is hardly in control. Correlatively, one can take down a dam, but one cannot take *back* a dam. The lives of people who live there will continue to change for decades, but in ways that cannot be controlled or even anticipated with much specificity. To stake one's future on a river is the beginning of a life lived in anticipation. To that extent, infrastructures define the limits and capacities of being human in fundamentally uncertain ways.

Over the past two decades, Laotians have participated in unique experiments in sustainable hydropower development. Attempting to deal with the many problems caused by large dams, these experiments have enrolled villagers into spirited interventions built around flexible trials of new livelihood strategies. Such programs of participatory inclusion and energetic trialing of many ideas to see what works have emerged out of American management theories that rely on personal motivation and an ethos open to the possibilities inherent in lived relationships. They contrast strongly with a view of planning built on foundational knowledge claims about social and natural relationships. In doing so, they foreground the crucial role of knowledge in configuring novel ecologies and reworking human capacities.

This focus on villagers' life practices, moreover, has occurred within broad public debates about resource-based development interventions for an industry whose development rationale has been seriously questioned since the 1980s. For a country with the potential for building many new dams and strong, popular development goals, Lao hydropower has attracted global attention for revitalizing an industry beleaguered by criticism (Singh 2009; Goldman 2005). The problems with large dams are well known, and they are no longer the temples of developmental modernity once hailed by Jawaharlal Nehru (Baviskar 2005; Khagram 2004; Sparke 2006; Farmer 1992). Dams displace people; they vitiate social cohesion and create poverty; they destroy ecosystems; they silt up and often cost more money than they are worth—and some are much worse than others (Scudder 2006). Here, a novel design of hydropower dams—smaller transbasin diversion schemes that generate comparable amounts of electricity from much smaller power facilities—has been linked up with a host of sustainability measures to deliver development benefits and minimize environmental damage to those living along these rivers. Enterprising work to remedy livelihood concerns has been accompanied by risk communication strategies including claims of sustainability achievement.

By studying the practices and reason of postnatural ecologies in situ, this research project opens up questions for environmental anthropology surrounding an entrepreneurial ethos bearing on human-environment interactions. Faced with the limited ability to understand, predict, or control socionatural relationships, especially under pressure from increasingly effective outside criticism, one response has been to develop forms of flexible experimentation on projects of life and living. Sustainability has taken form as a problematizing domain in which questions proliferate that often have no answers or in which the answers are inadequate but are nonetheless posed, considered, and experimented on. In this sense, sustainability must be understood as so many attempts to domesticate the excess of ecological events (on the event, see Deleuze 1990, and further discussion in the conclusion). Rather than a critique of sustainability representations, the ethnographic task is to understand the problematic situation as an active, affirmative dimension of life and living, including the vernacular and explicitly "interested" environmental knowledge that attempts to grapple with environmental events. When an ecological event provokes experimentation on how it is possible to live in such dynamic contexts, the result is not simply a degraded environment but rather a novel ecosystem in which new capacities for living are both engendered and disabled.[12] Hence, "anthropogenic" is a double relation: rivers are transformed and in turn transform people's lives in ways that demand experimentation on new capacities for living.

The common view is that anthropogenic causes of ecological degradation have resulted in a loss of biodiversity and ecological habitat. Degradation, viewed qualitatively and quantitatively, is a one-way, negative evaluation that does not account for the modes of affirmation that are essential both to capitalist production and to diverse programs of living in anthropogenic ecologies, nor does it account for the ways living beings, including people, contort themselves into cramped postures to accommodate the ecologies in question. The understanding presented in this book is that the active production of ecological uncertainty, or what we can call the postnatural ecologies of late industrialism (Fortun 2012b), represents not simply a degraded environment, but more aptly a situation of novel ecologies—new ecosystems that have their own conditions of vitality. Moreover, there is a strong predisposition to view the human agent as the static culprit of ecological harm or conversely the hapless victim of mute, disastrous forces. But "anthropogenic" goes both ways, and the figure suggests more dynamism than is commonly acknowledged. The entrepreneurial practices I describe in this book are spirited attempts to generate human projects of life and living to accommodate postnatural ecologies. Without implying any form of ecological determinism, anthropogenic rivers open up new potential for life and living, and this "hydrology of hope" (Hughes 2006) is attuned to new possibilities for living in uncertain environments.

Much of the inspiration for *Anthropogenic Rivers* comes from Hugh Raffles's ethnography *In Amazonia*, in which he writes, "The anthropogenic channels of Amazonia—great and small, old and new—are moments in the life of a region. Like grains of sand, their presence lies in their multiplicity; yet, they are also little gems, self-contained and faceted histories of counterpoint and relief. They signal other worlds, so normal, so commonplace, and everyday, yet also unfamiliar, and, in the persistence of that difference, relevatory of the boundaries that circumscribe inquiry" (2002, 42). Eben Kirksey argues for a view of novel environments most forcefully through his concept of emergent ecologies. Kirksey is concerned with actors who build out socionatural relations into new and frequently hopeful configurations. "Reaching into the future," he argues, "these thinkers and tinkerers are grabbing on to hopeful figures and bringing them into existence" (2015, 7). My research suggests that this kind of future-oriented experimentation has a deeper ethos that links it to a history of materialist practices that undergird the practical ontologies (Jensen and Morita 2015) of late capitalism. Beyond this historical context, ecological uncertainty also speaks to nondeterministic ways of understanding intertwined and mutually dependent ecological and social relations.

There is an emergent vitality and imaginative potential to ecological futures even when those futures appear bleak. For Nathan Sayre (2012), there is no easy way out of a long legacy of nature-culture dualisms, and the approach taken here is to inhabit the problematic of late industrial environments—to eschew a "critical politics of debunking" (Holbraad, Pedersen, Vivieros de Castro 2014) even regarding such suspect practices as late capitalist managerialism. The objective is show the river, turning again and again on itself through the multiplicity of relations through which it is enacted, as a manifold or plurality that becomes otherwise to itself, constantly. For Sayre, industrial capitalism has defined "the demise of the human-nature dualism and the tenacious hold it nonetheless maintains," marking a situation in which the dualism becomes "something unrecognizable, or uncognizable, in terms of our inherited concepts" (2012, 58). But there is something about late capitalist practices, specifically managerialism, that is amenable to apprehending reality as a materialist multiplicity in its own devious, politically motivated ways. Thinking uncertainty requires grappling with natures that always threaten to get out of hand. To "grapple" is to manage, from the Latin root *manus*, to engage with the river and its possibilities in an ongoing, open-ended manner, and to live with the thing indefinitely. The approach does not view managerial capitalism as the enemy but, as Holbraad, Pedersen, and Viveiros de Castro put it, "to pass through what we study . . . releasing shapes and forces that offer access to what might be called the dark side of things" (2014). Helen Verran (2012, 112) asks for anthropology to "develop familiarity" that "in good faith" can address generative, dangerous technologies. Even if frequently it cannot be answered, the question becomes *in what ways is it possible to live with anthropogenic ecologies?* In spite of itself, sustainability management has become one globally dominant, clearly insufficient, answer to this question.

The Theun-Hinboun Dam is a major hydropower installation in the Mekong River basin that, in response to a successful activist campaign, made important management and financial commitments to improving its environmental reputation. My research approach was to understand not just whether its work was sincere or effective, but whether a systematic engagement with environmental concern would change the practices and reason through which environments are known and enacted by industry actors. The company's initial response to activists was to develop a ten-year, US$10 million plan, in spite of the fact that it was not legally obliged to do so. Curious how transnational activists with no legal standing had achieved this feat, I organized my research to explore the practices through which newly articulated environmental concerns worked their way into industry operations.

This research extends and builds on important insights concerning the environments of late industrialism under conditions of risk and technological intervention.[13] In contrast to constructions of naturalized fact and authoritative expertise, the ethnographic data here show a different logic is at work. Environmental technicians and managers routinely acknowledge how little they know, even while they also obfuscate and defend their companies' failures. In many cases they claim that rigorous knowledge is too time-consuming and costly to form a basis for planning and that failure-prone planning itself places too much trust in knowledge.[14] In contrast they offer spirited experimentation with techniques, understood as a process of learning, with a diagnostic sense of understanding failures and rethinking ideas; this experimentation incorporates substantive negotiations over value and process with affected people. In the present case their work was also deeply opportunistic for the ways sustainability framed the company's reformed image and disrupted activists' assumptions. I do not claim that the process resulted in successful mitigation of environmental problems. The confluence of sustainability communication, widely uneven application of information-rich environmental monitoring practices, and techniques of risk management must be understood within a genealogy of information and environment as inevitably "open systems" (Fortun 2003). "The whole question," writes Ulrich Beck, "is how do we handle nature *after* it ends" (1999, 31, his emphasis).

Opportunism, like optimism, implies affirmation—and the affirmation of electricity cannot be reduced to the particularistic aims of corporate developers, no matter how much it depends on them. From the vantage of most middle-class Lao citizens (and indeed many rural agrarian actors), the affirmative, anticipatory values of electricity generation define the legitimacy of "Hydropower's Circle of Influence" (chapter 1). I ground Laos's hydrological project within its history of political marginalization, commitment to postsocialist "openness," and affirmation of anti-imperialist autonomy. Furthermore, making electricity is already an environmental act of stripping electrons from the gravity- and solar-fueled water cycle. Taking up electricity in terms of its broad social affirmation unsettles any easy polemic in favor of experimenting with how to live anthropogenically. Building a dam is an opportunistic achievement involving a diverse transnational ecology of investment and expertise, the results of which are highly uncertain. The lives of people who live there will continue to change for decades, but in ways that cannot be controlled or even anticipated with much specificity. The essential figure of such transformed ecologies is therefore one of temporalization, and the seductions of anthropogenic rivers give form to the contemporary Lao state.

Beginning in the mid-1990s, transnational antidam activism came to bear on new investments in Laos that were being touted as new kind of green design. In chapter 2, "Vulnerable at Every Joint," I reconstruct an important campaign that proved successful, contributing anthropological insight into environmental activism (e.g., Pearson 2009; Ottinger 2010; Welker 2009). Sanjeev Khagram (2002) argues that transnational activism "restructured" international politics in the 1990s, a phrasing that takes care not to overstate its influence.[15] In this chapter I explore the concept of technical entrepreneurialism and the closely allied term "political ontology." My interest lies in how environmental concerns became manifest practically and operationally within the industry. Activists destabilized planning networks and their cultures of expertise by demonstrating that hydropower's problems had not been taken seriously. This has greater significance than it seems at first blush because it implies that activists worked to create a problematization—indeed, to produce systemic uncertainty—rather than by constructing simply a different version of the facts or by convincing powerful agents to act more conscientiously. In doing so, they created a situation in which environment was no longer experienced as a vague context or external nature "out there" but as a kind of systemic vulnerability that could reappear at potentially any joint in the industry's complex institutional relations.

Organized by American theories of "management by results," managers' hands-off approach delegated practical responsibility to on-the-ground negotiations between Lao technicians and villagers (chapters 4 and 5, "The Ethics of Document Engineering" and "Anthropogenic Rivers"). Richard, the charismatic, performative American manager, played a "trickster" role in substantially reworking activists' practice into the company's own green communication strategy (chapter 3, "Performance-Based Management"). I describe how technicians and consultants struggle to determine the ethical dimensions of living along anthropogenic rivers and demonstrate the role of personal motivation, "results" thinking, and asymmetrical relations among Lao technicians and villagers necessary for the management by results approach to work. Affirmative values of regional developmental aspiration were integral to the work of dealing with environmental problems.[16] Projects based on flexibility, personal motivation, and experimentation reiterated the uncertainties surrounding the dam and, in the absence of long-term investments in knowledge and technique, failed to address the most debilitating problems of land access and food supply.

Management—essential for thinking about business relations to the environment—is a key site of the production of uncertainty and for understanding the anthropogenic double movement (chapters 3–5). Constituting a matrix of complex socionatural relations, management is a labor relationship to which unresolved, difficult to handle problems are delegated.[17] By viewing business activities as richly socionatural (rather than transparently self-interested or, on the other hand, strictly discursive), I show that considerable ethical and even political environmental concern routinely organizes commercial sustainability practices. Management practice constantly deploys materialist experimentation, including treating language and knowledge as durable things to be played with and undermined. At the same time, environmental management has vital implications and, like medical practice, involves "clinical" encounters through which normative claims are made and negotiated. To that extent, sustainability constitutes a biopolitics. Not confined to commercial objectives, management refers to a distinct domain of power-infused material practice that organizes intractable socionatural relations, and which in turn must extensively configure itself in terms of those relations. Furthermore, these practices surrounding anthropogenic rivers occur within a risk management approach in which flexibility and innovation routinely bleed into promissory rhetoric and communicative spin. Anthropogenic rivers form a domain in which "the imaginary and the real figure each other in concrete fact"—and, inevitably, in riparian flux (Haraway 1997, 2).

For the hydropower company, sustainability management required much more detailed work with villagers living along the rivers in question (chapter 5). Unlike environmental conservation efforts, sustainability promised better distribution of development goals and linked up nicely with regional affirmations of "prosperity" (*chaloen*).[18] A major economic effect of dams pertains to villagers unwilling to plan or invest in long-term life projects when faced with uncertain futures, while other villagers take advantage of new opportunities that promise windfall results. Jan Pieterse (2000) describes development as the management of a promise, and a promise is a kind of opportunistic speech act that takes advantage of the unstable possibilities at hand.

Stuart Kirsch argues that sustainability is a corporate oxymoron used to "conceal harm and neutralize dissent" (2010, 87; Benson and Kirsch 2010; Kirsch 2014). This view emphasizes the role of environmental risk in business practice, resulting transformed environments, and the mechanisms through which sustainability turns into opportunistic business practice. Yet it does not do enough to show the forms of "cruel optimism" through which villagers work to reimagine conditions of living with anthropogenic rivers (Berlant 2011). With its extensive experimentation, the entrepreneurial ethos troubles the idea that uncertainty can be reduced to a discursive or epistemic regime. The "small machines" (Biggs 2012) of sustainability—water pumps for gardens, hygienic pens for commercial pig stock—are compelling at some more intimate biopolitical and material level because they speculate on unstable possibilities and, in doing so, make risky commitments. Prone to failure and unintended consequences, the things themselves are uncertain.

Moreover, on its own, the discursive approach fails to adequately attend to active reinventions of nature (Banerjee 2003). Anna Tsing (2014, see also 2005) calls for "critical description" as an ethnographic mode that does not reduce ecological relations to the play of power or cultural meaning. To further develop the argument of the temporal structure of anthropogenic rivers, I show that changing temporalities of water flow result in new configurations of water and soil. Timing, predictability, and duration matter to the interplay among soil, water, and risk. The riparian zone is an open relation between wet and dry, taking form as unmitigated erosion and flooding, too much soil in the water, and land too waterlogged to farm. Anthropogenic rivers evoke what Kuntala Lahiri-Dutt calls "an intricate combination of earth and water, something that seems to be 'a land upon water or a watery sheet upon the land'" (2006, 394).[19] I document this interplay with respect to three dimensions: extensive riverbank erosion and waves of sand and silt that redefine riverbed geomorphology; prolonged flooding and "risky paddy"; and the unavailability of water, poor soil, and "permanent shifting cultivation." I describe villagers' ongoing negotiated relationships with environmental technicians who experiment with villagers' techniques, cajole them into working, explain to them the logic of the program, and grapple with excuses and explanations when things go wrong. The vitality of these novel natures—their risk and opportunity—is tied to villagers' affirmation of reworked modes of living.

Each chapter is written around a single materialization of environmental knowledge central to the action of the chapter. The chapters taken in sequence trace a progressively more detailed story about the river in question, starting from context-poor debates about hydropower and expertise that circulated in the nation's capital, continuing to discuss the activist campaign's minimal empirical characterization of the dam's impacts; practices of managing transnational environmentalism; the more detailed work of a team of experts; and finally the messy sustainability management of life along these changing rivers. Moreover, given that so many of the key expert figures are Anglophone expatriate men, an "authorized subtext" (so to speak) of the text is that of white masculinity and the fragile intimacies that emerge around the limits of expertise. From context-poor to context-rich, the chapters work to bring the reader closer to the river through a series of materialist negotiations and demonstrate the multilayered, transnational ecology in which anthropogenic rivers give form.

Late Industrial Environments

Late industrial environments can be identified as a class of environmental concerns forming a well-defined topic for anthropology of environment. I offer this problematization in distinction to the metanarrative of the Anthropocene, which I find to be teleological and too frequently to assume rather than problematize the status of the human in ecological contexts. Here I follow Kim Fortun's (2012b) argument concerning the ethnography of late industrialism. Fortun emphasizes the state of disrepair of industrial infrastructures, the slow disasters of toxic exposure and operational neglect, and the persistent failure of knowledge to provide even minimal understanding for maintaining life. The ecologies at stake are dirty, brown ecologies of deindustrialized urban landscapes, polluted waterways, and predictable but unpredicted nuclear accident zones. The 1984 Bhopal disaster—among many others, such as the 1986 Chernobyl meltdown or 1982 Chicago Tylenol poisonings—framed risk politics in Euro-American political discourse, coupling neoliberalism with specific ecological experiences of the demise of institutional guarantees of security. Risk politics presented a dilemma when public institutions were unable to guarantee the authority of expert knowledge on which public decisions could be made (Beck 1992; Callon, Lascoumes, and Barthe 2011; Funtowicz and Ravetz 1993). Risk discourse always had a Eurocentric bias because it was primarily within the logic of technocratic rationalism that the obsessive worry about foundational guarantees proved to be a disruptive weakness. The politics of risk hinged on the state's inability to act when it could not guarantee safety; its central figure is the precautionary approach, with its distinctly European flavor; or in the United States, a combination of pop culture, business bravado, environmental monitoring, and systemic injustice. Thirty years later, risk politics has shifted decisively toward an ontological politics in which temporalization and the production of uncertainty are critical features. Bruce Braun and Sarah Whatmore (2010, xxi) write that, "it is not enough just to say that things are lively and potent rather than dead or inert; rather, we wish to underline that things—and especially technological artifacts—carry with them a margin of indeterminacy. . . . Far from deterministic, technological artifacts temporalize, opening us to a future that we cannot fully appropriate even as they render us subject to a past that is not of our making."

This lack of determinism, what I call "underdetermination," forms a strong contrast with earlier environmental anthropology's interest in technological and environmental determinism and has the effect of introducing a temporality of imaginative investment into ecological relations themselves. Ecology, especially the dirty ecologies of late industrialism, is an imaginative domain insofar as it is constitutively underdetermined.[20] This condition of underdetermination *multiplies potential* and *temporalizes life*. Hence, uncertainty is an epistemic value that is built into socionatural relations, and those socionatural relations and entities are strictly speaking underdetermined. They are unfinished. Anthropogenic rivers force on subjects who are invested in those rivers the question of how things could be otherwise (see Povinelli 2011). What they are is not yet.

Late industrial environments are defined not by the truth-effects of power/knowledge or by the construction of durable scientific facts, but by the production of active, dynamic ecological events. These events are manifest in the unpredictability or instability of ecological relationships and the correlated insufficiency of knowledge or language to adequately apprehend, grasp, or contend with ecological uncertainties. Apprehension marks a condition of not knowing enough and an experience of anxiety. Late industrial environments are constituted by uncertainty.[21]

Integrating major contributions from the anthropology of science and technology, I suggest the ability to know and act on environments depends on long-term investment and accumulation of infrastructures of knowledge, practical technique, and regulatory control. Building something valuable requires an affirmation of consequences and long-term investment in the composition of durable relationships, including ideas. And yet there is pervasive underinvestment in the environmental dimensions of industrialism, especially concerning toxicology of chemical pollutants, the ecologies of genetically modified organisms, or the implications of radiation exposure. Discussing Chernobyl, Petryna 2002 shows how many layers of scientific claims, error, speculation, and explicit political manipulation of complex truths overlay and repeat the event, giving the aftermath its full historical and sociotechnical weight. Amita Baviskar (2005) describes the systemic livelihood uncertainties that result from hydropower development, a condition that emerges long before the dams are built. When environmental problems are saturated with power relations, a host of mechanisms serve to make these problems "imperceptible" (Murphy 2006), such as bureaucratic neglect, stakeholder management techniques, audit and sustainability reporting, and environmental monitoring. These are not mere discursive addenda but features of anthropogenic environments themselves, that is, conditions of life and living that become extensible through completely inorganic relations (such as information systems). Hence these are also the same mechanisms through which ecologies are known and dynamically acted on. Technical knowledge, its multiplicity of effects, and its inadequacies are built into the lived environment. Hence uncertainty has to be underscored as integral to environmental relations themselves, not simply a secondary epistemological consideration.

As Fortun writes, "understanding of the chemicals released in Bhopal remains inconclusive; they are among the over 100,000 chemicals registered with governments around the world for routine use; the data hasn't been collected, the science hasn't been done, to understand how these chemicals affect human and ecosystem health. Thousands of new chemicals continue to be introduced each year" (2012b, 446). This is not simply a statement of fact, but rather it marks a temporality of threat. This disjuncture between manifold ecological events and what is known and understood about them is expressed in varying degrees for large hydropower dams, genetically modified organisms, anthropogenic carbon emissions, geoengineering, large-scale mining and deforestation, and biodiversity loss on epochal timescales. Knowledge of how to live and the language available for living are rendered inadequate with vital consequences, what Fortun calls discursive risk, in which "misrecognition of what is going on . . . effectively quells the possibility of a different kind of future" (2012a, 326n17).

Michelle Murphy writes of the "temporal aspect of chemical infrastructures: their slowness, their persistence, their creeping accumulation, their latency. . . . Latency in ecological time describes how the submerged sediments of the past arrive in the present to disrupt the reproduction of the same. Through latency, the future is already altered" (2013, 106). Echoing Karen Barad's 2003 work on performativity and materialization, Murphy argues that toxic "exposures are made to matter" (2006, 18), rendering a historical ontology in which the specificity of anthropogenic relations take form. This material specificity consists not in the obvious facticity of things, but in toxicity's imperceptibility, within histories that tell "how chemical exposures became quintessentially uncertain events" (14). The ontology of late industrial environments is "materialized as *uncertain*" (7; her emphasis), with vital consequences. For instance, environmental laws cannot be enforced when the knowledge infrastructure for demonstrating legal standards of proof does not exist, making clear that environmental justice can sometimes boil down simply to being able to enforce existing legal standards. Questions of temporality also impinge on the modes of stasis essential to late liberal ways of governing ecology. Murphy (2013, 2006) argues that questions of toxicity and exposure have been routinely managed through monitoring rather than regulation, while Fortun (2012a) describes the informating of environment as directed at communities of color and other marginalized groups unprotected by regulation.

The temporality of threat is integral to affirmations of life and living. As a conceptual move, uncertainty brings together critical science studies, environmental justice, and biopolitical analysis of historical power relations. Genealogies of knowledge/power are reconfigured to focus on the contingency of historical achievements and capacities rather than the apparent naturalness or solidity of power's truth effects. Denaturalization strategies, including those of Latourian actor network theory, remain central, but when critique forms the end point of analysis (Riles 2006) without a concomitant affirmation, it neglects the manifold ways in which power works to destabilize natural relations, produce epistemic uncertainty, or systemically defund or delegitimize certain projects of knowledge. It also absolves the critic of the difficult labor of speculative composition that is at the core of all affirmative politics. The fact that working with things takes time and is inherently contingent already inculcates an ontological principle, namely that material-semiotic relations form the very capacity and limit for imaginative possibility. Critique must ask itself what inscribes the limit of its imagination.

Post–actor network theory and renascent ontological approaches seek to reconstruct network ecologies and assemblages of various kinds in their own move to dismantle nature-culture distinctions, to move away from an ontology of categories (nature, society, culture, science), and to affirm "relations" as a basic metaphysical term. My work remains very close to those methodological positions. Yet many actor network constructions betray a profound difficulty with questions of value, largely out of reticence of over-privileging the human subject. Indeed, one feature of posthuman environmental critique is its wariness of environmental justice and postcolonial concerns that are less eager to embrace technological futures "whatever the risks, whatever the as-yet-unforeseen, mutagenic transformations of nature-culture, be they happenstance or engineered" as Lawrence Buell has recently argued (2011, xiv). But as Elizabeth Povinelli (2016) shows, ontological politics is not a liberal version of multicultural difference but closer to the black liberation rallying cry, "I can't breathe." Racial justice and environmental justice critiques hinge on the fact that certain human lives are explicitly devalued, a claim that demands a critical historicization of the European concept of the human and its highly political application (for a biopolitical analysis, see Weheliye 2014).

The history of devaluing the lives of certain groups of people thus needs to inform and flush out the massive devaluation of all sorts of socionatural relations (Merchant 1980). There is no reason to view value, value-making, or practices of evaluation as the exclusive product of human subjectivity. Living beings judge, affirm, and reject the enabling and harming relations in which they are enmeshed. By the same token, ecological relations imply knowledge beyond the human in some form, or rather "knowledge" becomes a generalized question that forces us to look at nonhuman living beings in fundamentally different ways and even to ask what is meant by the concept (e.g., Myers 2015; Kohn 2013; Viveiros de Castro 2014; Haraway 2003). Reciprocally, if being human is rejected as a foundational project or a matter of establishing a category of biological, spiritual, or metaphysical uniqueness then it can better become a site, as Anand Pandian (forthcoming) claims, for creative emergence of a "humanity yet to come." It is, above all, the *question* of the human that forms a critical component of ecological thought today.

The anthropology in question here is thus a kind of speculative anthropology—a detailed study of one set of possibilities for the anthropogenic, in which the human is taken as a question and an inevitably historical potential in need of political reworking. What is being formulated in practice, and deserves to be investigated ethnographically, is a range of historically novel relations for grappling with the underdetermined potential of late industrial environments. The explicitly valued temporalities of threat, opportunity, latency, and even mourning underscore *harm* not merely as an effect but as a deeply historicized process in which temporality and subjectivation are implicated in novel biopolitical environments. To become a harmed body means becoming a person capable of bearing harm—a person vitally embodying the afterlives of late industrialism. That capacity is powerful in its own right.

Interlude

On the Postcolony (Engineering)

A French-Canadian man, one of the engineers I worked with in Laos, was a lead construction engineer for the Martyrs' Monument in Algiers. Martyrs' Monument was a pet project of the Canadian prime minister at the time, Raymond told me, a gift from the people of Canada to the newly independent Algeria. He described that project as what really got him hooked on international construction work. "You know these architects," he said, "They want you to build a dream."

With his hands, Raymond mimicked the contours of the hundred-meter column, which came up from a curved tripod base only to split into three again at the apex. At the base is a war museum and at the top, a meditation space. A seductive shape. One of the challenges he faced was to route concrete trucks from the port where the concrete was mixed on barges, through downtown traffic, to be poured on a precise schedule. This is the sort of work construction engineers do, different from design engineers.

The Canadian prime minister was intent that the monument be starkly white, as a symbol of peace and independence and in keeping with the city's Mediterranean aesthetic.

However, as it turned out, the base limestone for the concrete was nearly black. It is possible to bleach and color concrete, but not by much. Of course, politicians think anything merely technical can be done, so the engineer had to convince the administration that pure white was in fact impossible. The prime minister was to visit Algiers and make a decision; the engineer prepared an array of some twenty pigmented samples to demonstrate the possible. The engineer was not allowed to be in the room when the decision was made; based on the samples, the prime minister decided on an off-white dye for the concrete. The project won an award for innovation in construction techniques.

In the telling, the engineer revealed much about the affective values of expertise and the pathos of large-scale nationalist construction projects. He grasped for the monument's sublimity, appropriate to the nostalgic drama of nationalist iconography. Awaiting the politician's decision from the next room, the aesthetic symbolism gifted to the postcolony was at stake. This marked the very social organization of technopolitical expertise—the politician on one side, the engineer on the other. The engineer's task was to demonstrate the potential of the dark limestone, but he was also the mediator of possibility and constraint as it bore on the high modernist dreams of architects and politicians. He offered the figure of the engineer as a kind of hero, not disproportionate to Nehru's invocation of hydropower dams as the temples of modern India.

Chapter 1

Hydropower's Circle of Influence

In the late 1990s and early 2000s, Laos was a crucial test case globally for revitalizing hydropower as a viable option for international development. Through the establishment of loan conditionalities, writing of new environmental laws, and attempts at "soft" programs of structural adjustment by the World Bank, Asian Development Bank, and International Monetary Fund (IMF), Lao hydropower came to form an exemplar for reworking the ability of multilateral lenders to finance large dams under the rubric of sustainability. For international lenders and many foreign development workers and NGO staff, involvement in Lao hydropower revolved around attempts to reform the Lao state, at least partly by institutionalizing global environmentalism. At the same time, for Lao officials and an emerging urban technical class, large dams were an integral part of an affirmation of national development in response to the country's postsocialist predicament of long-standing isolation and geopolitical marginalization.[1]

Viewed from the perspective of either international development organizations or Vientiane's emerging bureaucratic and technical class, sustainable hydropower bore directly on the contemporary form of the Lao state, albeit in strikingly different ways. By contrast, from the vantage of the anthropogenic river, it was clear that, by the end of the 1990s, Lao political economy had thoroughly molded itself around the opportunism, promises, and even hype of its hydrological and energetic potential. How have rivers come to saturate and at times even inundate Lao postsocialist developmentalism? If we are not going to allow the engineers, financiers, and developers mastery over rivers, then it is important to retheorize the historically specific modes of subjectivity whereby agentive nonhuman relations come to maintain such a striking presence in contemporary political infrastructures. Anthropogenic rivers, I show in this chapter, operate as an imaginative and seductive potential in Laos's postsocialist political affairs.

If the electricity grid is an "organization of enabling power that allows any invention of statecraft to occur in the first place" (Boyer 2015, 533), the Lao grid has been built for export—and its politics follows. Hydropower's vast material claims promised independence and prosperity to address the country's history of political marginalization, war, and isolation, but they did so in part by compromising the ideal of national sovereignty. The liberalization of investment rules in the late 1980s brought a rapid escalation in big dam projects, fueled by new contractual tools and concessionary arrangements that distributed the governance of ecological events among diverse nonstate partners. As Aihwa Ong (2006, 8) argues, "complex interactions between diverse zones and particular networks [of capital] challenge sweeping claims about a unified landscape" of neoliberalism. The sheer range of resource investments—coupled with the explosion of development aid projects iconic of Lao developmentalism—created a highly variegated, sometimes chaotic terrain in which foreign actors are involved in spatially circumscribed, heterogeneous regimes of investment and rule. Postsocialist Laos has been governed both by continued commitment to centralist, single-party rule and a deliberate policy of openness toward collaborations with transnational actors, whether commercial, NGO, or governmental. These collaborations have a range of diverse governmental objectives; they often demonstrate norms of capitalist accumulation but are rarely reducible to them; and they routinely articulate sustainable development objectives, even if these are rarely achieved in practice.

In contrast to concern with short-term efforts at reforming the Lao state in line with global environmental norms (Goldman 2005), I emphasize the longer-term postsocialist program of managing novel articulations of capital investment in light of stabilized programs of governance and national autonomy. I use the term "sustainability enclave" to describe the territorialization of hydropower under broadly neoliberal conditions of investment and rule.[2] This mode of concessionary rule—what Ong in part calls neoliberal exceptionalism—was both a shrewd strategy whereby the Lao state was able to manage its incredibly complex postsocialist predicament and a kind of "private indirect government" (Mbembe 2001; Ferguson 2006) or multiplicity of concessions in which sustainability demands could be articulated but only in highly variegated ways. Conversely, Lao postsocialism is an outcome of anthropogenic rivers' creative potential insofar as its reforms are premised and promised upon unique riparian configurations. In many respects, Laos's postsocialist geopolitical predicament, including its distinct forms of national expertise, can be seen as an outcome of rivers' anthropogenic potential.

Late Industrial Rivers

During the decade of this research, 2001–2010, seven major hydropower dams were built in Laos. In the previous decade, the first of Laos's postsocialist constitution and open economy, at least two major projects came into operation—as well as a small domestic project that subsequently collapsed during a tropical storm. One project was built in the 1980s. In the early 1970s, at the height of fighting during the Second Indochina War, the Nam Ngum Dam was completed with American financing and Japanese construction—in a sense, it is a symbol of the geopolitical crucible that formed modern Laos. But from 2010 until the end of 2014, some twenty-five new dam projects were either completed or under construction across the country (figure 1.1). With nearly all of the electricity in question being exported, Thailand's rapid industrialization was essential to the industrial transformation of Laos's rivers. The rapid rise in energy generation and the resulting hive of activity bearing on Laos's rivers have given practical meaning to the national policy, aspiration, and marketing slogan of turning Laos into the "battery of Southeast Asia" (Ferrie 2010).

This construction binge was all the more striking because the Mekong River basin remained largely free of large dams until the 2000s. There are exceptions, of course. International Rivers 2013 lists six completed mainstream

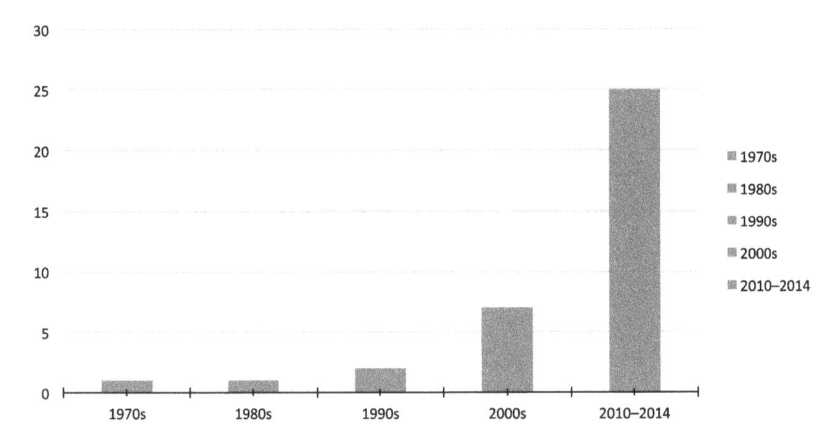

Figure 1.1. Major Lao hydropower projects by decade. Preconstruction and projects fifty megawatts and smaller excluded. Sources: International Rivers 2010; CGIAR 2013.

dams in China, where the Mekong is called the Lancang River—a reference to the historical kingdom of Laos, Lanxang. But what made Lao hydropower so distinctive was the rush to build a large number of much smaller and technologically distinctive dams along the country's ecologically sensitive Mekong tributaries.

If Lao rivers have such powers of attraction, it is because hydropower promises an event. A brief vignette can introduce some of this dynamism. The Xe Bang Fai River is a prominent tributary of the Mekong. It drains a substantial watershed to the west of the Annamite Cordillera, the narrow, well-defined mountain ridge that forms the border with Vietnam—the southern reaches of which formed the terrain of the Ho Chi Minh Trail. Lao filmmaker Maiveng has collaborated with Mekong Watch, a Tokyo-based environmental NGO, to document the cultural ecology of the Xe Bang Fai River. In 2004, I joined him and some North American activists in a modest apartment in Vientiane for a private screening of his film, which had been produced for Lao broadcast television. Imagery of blue-green waters and vibrant rapids inform ethnological footage of virtuoso fishing techniques and the riparian ecology of village life. One village-based welfare and river protection measure, which they documented, involved prohibition of fishing along a stretch of the river; poor families were allowed to fish in this rich area provided they pay a small fee to the village fund. It is an idyllic portrait.

However, the target of Maiveng's film is somewhat different than a simple celebration of life along the river. The Xe Bang Fai is the drainage for Laos's largest hydropower complex, which became operational in 2010. The project injects up to 315 m^3/s additional water into the river, diverted from the higher elevation, adjacent Theun River and effectively doubling the flow of the Xe Bang Fai for a large stretch of the river. Ian Baird, Bruce Shoemaker, and Kanokwan Manorom (2015) document systemic change due to erosion, change in aquatic resources such as plant life, and contraction of the fisheries. While many of these changes are effectively invisible and the riparian villages seem to be the very paradigm of undeveloped rural life, the agrarian and aquatic ecology is effectively dominated by an industrial installation.

When one foreign activist at the screening pushed Maiveng to explain why, in his beautifully shot documentary of a river at the center of a vital environmental campaign, the film failed to include any reference to the hydropower dam, he replied that he wanted to create a positive view of the vitality of life along the river and that the documentary was distinguished for being shown widely on government-run television. Another activist at the screening, an American, pointed out that he might be arrested or suppressed if he tried to make the politics explicit, but that was not the filmmaker's point. In a situation in which people already knew the context, the meaning was already clear without the need for confrontation. Difficult to recognize from the vantage of Western-style political confrontation, this style of critical affirmation is an important element of Lao development politics.

The industrialization of a hydrological landscape composes these rivers as industrial drainage and, effectively, involves a partial evacuation of life from their waters. As Baird, Shoemaker, and Manorom (2015, 1090) report, "due to the decline of the [Xe Bang Fai] fishery, [a major fish] market no longer exists. Villagers reported that the sale of labor (including through illegal migration to Thailand) is now their most important source of income. There is not even sufficient fish for local consumption—villagers report having to buy canned fish from Thailand." The late industrial landscape demonstrates an open-ended temporality that thrusts people into new power relations. Here the objective is to understand the national frame in which dozens of these projects proliferate within the large-scale development of anthropogenic rivers. The immediate historical context is the hydrological planning of the US imperial project in the greater Mekong region during the 1960s–70s.

Development and War as Parallel Strategies in Southeast Asia

As Theresa Wong (2010) argues, the current emphasis on Mekong hydrological infrastructure must be framed against the vast planning endeavor of the US-backed Mekong Committee from 1957 to 1971. Cast against a history of equal parts marginalization, isolation, and dependency, the current condition of development aid (*leua* [excess, remainder, leftovers]) marks a history of subordination from which hydropower promises some relief, in form of participation on equal footing. More broadly, the postsocialist condition of Lao developmentalism has to be understood against the historical role of hydropower development in the regional context of overlapping, competing imperial ventures. The systematic destabilization of newly independent Laos during the 1950s and 1960s powerfully demonstrated the country's precarious dependence on far larger geopolitical allies and set the stage for a cautious but deliberate policy of managing foreign investment when the socialist project would begin to collapse in 1986.

"Laos in 1968 was a laboratory for economic development and modernization," wrote Paul Langer for the Rand Corporation, echoing the colonial "laboratory" of French urbanism identified by Paul Rabinow (Langer 1969, 7; Rabinow 1989, 289). On the contrary, the country was primarily a major theater of war after years of active geopolitical destabilization. Water engineering was essential to American involvement in the Mekong River basin during and prior to the Second Indochina War (Biggs 2006, 2012; Nguyen 1999). The Mekong Committee was the central international body in force, the predecessor to today's Mekong River Commission, which produced a major basin-wide planning and technical assistance effort in the late 1950s–1960s advocating the total transformation of the river for irrigation, power, and transportation access for moderately sized ocean going vessels. Whereas it is often described as being under the auspices of the neutral United Nations ECAFE (Wong 2010, 55; White 1963), Thi Dieu Nguyen shows that the Mekong Committee emerged from the highly partisan Colombo plan, which had a multilateral character but was actually bilateral in operation with the United States providing 90 percent of the funding under its Mutual Security Program (Nguyen 1999, 65–66). Nguyen describes the Mekong Committee as "nothing more than an intermediary," whose legitimacy stemmed from being a screen for US interests (67–68). Its primary role was technical aid in the form of planning, field studies, and data collection, which were all part of US attempts to establish its Cold War cultural hegemony in the region. The Mekong River was central to its imperial imagination.

American cultures of materialism of this imperial venture are instructive. David Biggs describes the US Bureau of Reclamation's involvement as instilling a culture of engineering expertise through technical advising, the promotion of modernization ideologies, and active training programs for Asian engineers. Referring to Thailand's massive Bhumipol Dam, he writes, "The dam, with all of its complex generating equipment and design features, not only consolidated American forms into the local environment but [also] simultaneously opened up new ways of organizing Thai engineers, water users, bureaucrats, and laborers" (2006, 236). The wholesale regulation of the Mekong was certainly predicated on a rationality of human control over nature, but it is important to emphasize the dimensions of freedom invested in technological and material relations within the American project. As a template for anticommunist development in Asia, David Lilienthal's active promotion of the Tennessee Valley Authority emphasized the active role of technical and material relations in people's lives, what he called "a language of things close to the lives of people" (cited in Wong 2010, 52).

The Mekong Project's director, Gilbert White, writing in *Scientific American*, also notes "the values of early experience and action" (1963, 59) in the development of Asian expertise, even if it meant a less than optimum project. Similarly, development theorist Albert Hirschman ([1967] 1995) believed that difficult development projects, appearing easy or straight-forward at the outset, provided one of the main benefits of international aid when local practitioners were forced to solve seemingly insurmountable obstacles. What Hirschman called the "hiding hand of ignorance" allowed for the risk averse to embark on "the long voyage of discovery in the most varied domains, from technology to politics" ([1967] 1995, 35). As White put it, "in setting human and cultural demands of this kind, the program for water management in the Mekong Basin becomes the organizing force behind more subtle aspects of social change" (1963, 58). Infrastructure implied obligation within a culture of didactic materialism; the Mekong would be Southeast Asians' teacher.

Such a midcentury culture of expertise and its material commitments was much less established in Laos, but it is clear the engineering regime was not uniquely American. Laos's major project during the Second Indochina War, the Nam Ngum, was built at the edge of the Vientiane Plain precisely where the capital city was most vulnerable to the communist Pathet Lao. Financed by Americans along with grants from seven other countries in the West (White 1963; Nguyen 1999, 126), the project was built by Nippon Koei under the leadership of Japanese wartime engineer Kubota Yutaka, who was already famous for building the world's then-largest hydropower facility in the Yalu River Basin in Korea during 1930–1945 (Moore 2013b). It was Kubota who drew up the blueprints for Mekong River's integrated hydropower development for the Mekong Committee. Aaron Moore 2014 argues that Kubota's Nippon Koei was "involved [in] the larger post-colonial security system in Asia directed by the United States" and emphasizes Japan's colonial and imperial legacy in Kubota's efforts to establish Southeast Asia as a market for Japanese manufacturing and expertise through subsidy systems based on donor aid and reparations. While Japanese philosophies of technology emphasized both comprehensive rational planning and the dynamism of art and technique (Moore 2013a, 28), it is not clear whether Japanese technical aid ever had the proselytizing elements of America's technical materialism.

Imperial infrastructure of the Mekong border zone and its associated knowledges gained a powerful affective element in the security context of the Cold War, much in the way Thongchai Winichakul (1994) shows that anticommunist fervor in Thailand arose in part through the material practices of territorial mapping. Development and war can be viewed as parallel strategies of US involvement in Southeast Asia. Both are addressed to the same problematization, that is, keeping communism at bay in the context of rapid decolonization. Such dynamic and unpredictable compound events of world war, decolonization, and political crisis made hydropower dams plausible security interventions. In a similar vein, David Milne (2007) shows that Rostow's commitment to industrialization within successive, universal stages of capitalist growth—an explicitly anticommunist agenda—also forged the basis of his persistent advice that the United States should aggressively bomb North Vietnam's fledgling industrial base. "Walt Rostow adhered rigidly to a linear theory of economic development. This theory informed both his passionate advocacy of foreign development aid and his vehement calls for the heavy bombing of North Vietnam" (Milne 2007, 203). Sharing their politics but on opposite ends of the infrastructural spectrum, building dams and dropping bombs were parallel strategies of American dominance in Southeast Asia.

One the other hand, Laos's independence and neutrality became impossible to maintain on this hot front of the Cold War. America's powerful sphere of influence over postcolonial Laos must be juxtaposed against North Vietnam's increasingly important influence over and support for the Lao communist insurgency. As White described it, "given the political history of the region during this period, the continuation of the [Mekong Committee's] work seems incredible" (1963, 49). The Vietnamese-backed communist insurgency had allowed for construction to proceed on Nam Ngum—a savvy but risky move considering it was never guaranteed they would take control over that very significant piece of infrastructure (Nguyen 1999, 125). During the rise of geopolitical polarization in the years following Laos's independence from France in 1954, it was the country's neutrality that became increasingly untenable. The United States did all it could to undermine the Lao government's program of neutrality, while the Democratic Republic of Vietnam in response actively supported and instructed the insurgency (Stuart-Fox 1997, 104). As the American military project collapsed and the success of the revolution became clear, the Mekong River, forming the now-closed border with Thailand, became one of the starkest manifestations of the country's marginalization and dependency amid hostile competing imperialisms.

Sustainability Enclaves

Against the backdrop of decades of destabilization, polarization, and war, the contemporary affirmation of large dams under concrete policy reforms associated with neoliberalism must be understood as a high-risk strategy of opening up a society and an economy to an unpredictable program of foreign investment and rule that inaugurated export-oriented power generation as a policy of statecraft.[3] This process of achieving "openness" involved a profound tension between openness to foreign investment, international aid, and cultural influence, and the attempt to maintain political consolidation and national unity of a dispersed, hard-to-govern territory (Jerndal and Rigg 1998, 824). If "neoliberalism" was at its most triumphal during the ideological movements of the late 1980s and early 1990s, from the vantage of the capital city, Laos's postsocialist experiment involved an important attempt to reenter the flux and play of regional geopolitics historically dominated by the threat of imperial forces.

Lao hydropower participates in this "propagation of freedom" through managing access of foreign investors to resources and the risks essential to neoliberal transformations of nature. The sustainability enclave is an emergent form of environmental governmentality that helps explain how largely foreign management of environmental matters has come into force in the context of Laos's postsocialist predicament. I understand neoliberalism as a complex of intensified modes of accumulation combined with heterogeneous strategies of governing freedom (Ong 2006; Rose 1999; Harvey 2011; Foucault 2008; Swyngedouw 2007; Kirsch 2014; Johns 2015). Cori Hayden 2003 demonstrates how bioprospecting, or the attempt to capitalize pharmaceutically active compounds from biodiverse ecologies of Mexico, has modeled benefit-sharing contracts on North American Free Trade Agreement–style free trade arrangements, while incorporating participatory programs for interpellating rural Mexicans into the role of conservation stewards of these damaged, threatened, and potentially enhanced anthropogenic environments. Like Ong's research on free trade zones and globalized relations of production (1987, 2006), Hayden shows the extent to which specific legal and extraterritorial configurations matter to new assemblages of capital, life, and labor. Practices of governing capitalist extraction *and* foreign development intervention are essential to Laos's strategic courting of resource investment; sustainability enclaves help to explain how this happens as an emergent form.

For neoliberal critics of development in the 1980s, hydropower was an important target for attempts to roll back financing of large state-led investments. Classically, hydropower was almost exclusively a state-centered development intervention, at least in developing countries, requiring large development bank loans and direct state control over all the main project elements such as construction risk, energy demand risk, hydrological risk, and so forth (Ribeiro 1994). High-risk dams have frequently made for bad public investments and have been paradigmatic of the critiques of statist development planning. Subject to mismanagement, cost overruns, poorly understood hydrological and geological factors, and manipulation of demand forecasts, large dams were often criticized by neoliberal conservatives and environmental activists alike as wrong-headed. In that context, dams built or commissioned in Laos in the 1990s became textbook cases for innovative public-private financing (Head 2000; Lee 2006) and received industry rewards for completion ahead of schedule and under budget.

Lao rivers were doubly anthropogenic, for the innovation of financial, technical, and environmental forms deftly sought to capitalize on the unique opportunities of Laos's hydrological landscape, while, at the same time, that hydrological landscape was becoming powerfully transformed. The crucial financial feature was a novel form of risk-sharing contract that allowed for greatly enhanced public-private joint ventures with foreign investors. Known as BOOT contracts, these financial arrangements let the state participate as investment partner while subjecting the projects to the extensive risk analysis of private financial, technical, and institutional investors.[4] BOOT stands for build, own, operate, transfer: it encapsulates a model of public infrastructure development that allows for a concession, usually twenty-five to forty years, in which a consortium develops, owns, and operates a power facility before transferring ownership to the sovereign party. The contractual mechanism both opened Lao resources to capital intensive foreign project investment and allowed for those investors to take a far more risk-centric approach to project outcomes. These concessionary forms allowed for the state to be a partial investor in a private project, including the collection of project revenue for energy exports, while also collecting royalties and taxes from the concession and energy sales, respectively. Concessions manage project risks related to resource development.

Speculative project acquisition in the early 1990s involved a rush to lay claim to concessions and to maximize the potential payout, sometimes by deliberately exploiting planning uncertainties. The idea was to flip projects when serious investors emerged. For instance, one major project known as Nam Theun 3 (later developed as the Theun-Hinboun Expansion Project) was initially planned by Snowy Mountain Engineering Corporation in 1991, under a concession agreement that required Snowy Mountain Engineering Corporation to initiate construction within ten years. I later worked closely with engineers, environmental scientists, and project managers of a different company when this project was being put to construction beginning in 2004. At that time, it was discovered that the initial project design grossly overestimated the potential volume of the reservoir. Not only did different maps show widely divergent elevations for the riverbed, but project developers were unpleasantly surprised to realize that the original Snowy Mountain Engineering Corporation engineers apparently deliberately used those maps showing the greatest reservoir volume and hence the greatest value of the project (see the interlude "What Is a Dam?"). More than simply the sociotechnical construction of facts, the deliberate play of constitutive uncertainty and speculative anticipation defined some of the early "prospecting" feel of Lao hydropower in the 1990s. However, projects purchased to be flipped were exceptions, and much of the work during that time involved coordinating BOOT-led private-sector investment into serious projects with multilateral support.

Underwritten by BOOT contractual arrangements, hydropower concessions have a definite spatial element that can be viewed as part of a deliberate, knowledgeable state strategy of managing relations with powerful foreign parties that themselves have divergent interests. The terms of these negotiated relations are often clearly asymmetrical, but it cannot be said that the prerogatives of Lao people, even disenfranchised rural populations, are simply displaced by foreign and elite objectives (see especially Dwyer 2013, 2014a). As Miles Kenney-Lazar 2010 notes, it is also important to better understand how specific populations can be sometimes wholly neglected within the chaotic, insufficient, and often experimental aspects of rule in Laos—questions that will come to the fore in chapter 5 (see also Baird 2014). Understanding what constitutes the scope of government is crucial, and care has to be taken to understand the specifics of Laos's form of paternal postsocialism in the context of affirmative demands made on government. Laos's sustainability enclaves are ad hoc and exceptional, built with discrete, usually transnational, partnerships around specific development or investment projects (Whitington 2012; on similar dynamics in African conservation see Ferguson 2006, chap. 1). The projects—ranging from limited NGO efforts to dams and large-scale agricultural schemes—are spatially and temporally circumscribed and may be short-lived and superficial. Alliances with diverse foreign actors in Laos have created a variegated landscape of projects that are territorially extensive, numerous, and flexible in their execution.

These formalizations and ad hoc arrangements demonstrate the postsocialist, developmentalist predicament of playing host to a vast array of widely differing projects and programs of national improvement. Hydropower and other kinds of resource projects increasingly can be viewed as sustainability enclaves when attempts to establish social and environmental protections take the form of a proliferating multiplicity of developmental techniques. Sustainability refers not simply to environmental protection, but to questions about how economic growth happens, with special attention to complex articulations of ecology and economy for a country heavily dependent on agriculture and fisheries for its base economic safety net. Clearly, the term "sustainability" does not refer to actual outcomes or to sincere intentions. Rather, it helps organize a flexible repertoire of practices that lead to outcomes beyond anyone's goals or interests; the enclave effect is one of those outcomes.

Aihwa Ong (2000) uses the term "graduated sovereignty" to refer to specialized zones of production enabled through customized legal and practical arrangements in Southeast Asia. Sustainability enclaves similarly involve routine, more or less deliberate delegation of problems of environmental and social management to foreign-run companies. While involving similar, important reworkings of sovereignty, sustainability enclaves display crucial differences. First, whereas graduated sovereignty relies on a stable form (free trade zones, special infrastructure packages) to court diverse manufacturing activity, the concessionary form of Lao political organization manages all kinds of foreign operations in which the prerogatives and practice government are explicitly being reworked. Second, whereas the globalized postdevelopmentalism relevant for Ong's argument is more narrowly economic and managerial, the political form of a multiplicity of overlapping regimes enables highly diverse normative agendas. Finally, hydropower companies routinely do the work of managing environmental and social issues through parastate or private indirect governmental arrangements. In the late 1990s, it was surely the ideal of foreign investors to let the Lao government handle social and environmental issues, but it turned out differently (see chapter 2).

How private developers come to take up these forms of private indirect government is part of the highly variegated territoriality characteristic of Lao postsocialist developmentalism. In 2008, the Nam Ngum 5 hydropower project, negotiated directly with the Chinese government and undertaken by Sino-hydro, China's largest water infrastructure construction firm, opportunistically applied for political risk insurance from the World Bank's Multilateral Investment Guarantee Agency (MIGA). To qualify for this insurance, the project must meet the World Bank's environmental safeguard standards. Intriguingly, both the MIGA officer and the Sino-hydro representative I interviewed said that the environmental requirements positively influenced the decision to seek MIGA insurance. The reason was that Sino-hydro felt it would be an easy project to mitigate, and they desperately needed to improve their environmental reputation. Furthermore, Sino-hydro has a history of serious construction and political setbacks for the many bilateral diplomacy projects the Chinese government demands of them and exorbitant rates for political risk insurance from Sinosure, the Chinese export bank's insurance arm. Their regional engineering director, based in Bangkok, had many questions for me about BOOT risk-sharing agreements that are otherwise the norm for hydro-development in Laos. Sustainability took on the form of a concrete but opportunistic financial risk-management mechanism depending on all sorts of different factors between reputational, financial, and environmental outcomes. And it formed an ad hoc governmental arrangement, in this case, for only 280 households in four villages in a remote area of Laos's highlands.

Through sustainability concerns, Lao rivers proliferate a plurality of relationships and investments, financial and otherwise. During the public hearings for this project, approximately a dozen villagers who were slated to be resettled spoke of their profound desire to be moved to the lowlands where they could be trained for new jobs. There was no question about the sincerity of their expression. But they interpreted the meeting as a negotiating forum (e.g., "on these conditions we will accept resettlement"), whereas others seemed to interpret it either as a performance of assent within a state ritual or as something more like confirmation of informed consent. Across from them, while they spoke, two officials sat prominently smirking and talking to themselves—a powerful reminder of the performative context of the meeting, the intricate web of communist party relations in which village politics are enmeshed, and the suppression of riparian lives. In fact, what farmers hoped for was not in the offering. Later, the MIGA representative demurred from speculating on whether they had a clear understanding of the situation, although in the meeting he tried to ask how they understood the pending livelihood changes. Like the work of filmmaker Maiveng and the dynamic engagement with foreign activists, these kinds of "thin" engagements cannot really be written off as epiphenomenal. They are surely sideshows to the main events of livelihood transformation and financial maneuvering, but they are also enactments of environment in their own right and part of the open-ended temporalities of anthropogenic rivers.

In a rare public glimpse of the production uncertainty that characterizes the ecological effects of dams, during the public hearing, an official with the Ministry of Agriculture blankly asked the villagers the only really hard question, that is, how they planned to eat after their paddy fields were submerged. They had no answer.

Laos's rivers through the 1990s and 2000s were subject to experimentation and novel forms of financial investment, even while the country sought to establish hydropower as a central element of its participation in international financial and governance networks. Classically, hydropower was a form of national development for which project risk was borne almost solely by states, and environmental damage was suffered by affected people. States were liable for planning outcomes, vouched for dams' economic benefit, and underwrote investment risks; states clamped down on dissent and ensured bureaucratized forms of violence and disenfranchisement when it came to managing environmental effects. When projects were not financially viable, it was national governments that paid the bill. The neoliberalization of hydropower development attracted a highly diverse range of actors and expanded possibilities for project design and environmental engagement through a process that involved a total reworking of the contours of risk within the industry, what legal anthropologist Fleur Johns (2015) calls a process of "failing forward." That failing forward included the subjectivation of ecological risk, at least in part. This subjectivation of risk also involved reworking the territoriality of the sovereign state through the concessionary status of investment and experimentation surrounding environmental sustainability. Preeminent among this experimentation was the World Bank's neoliberal structural reform program that, like the twinned crucible of war and development, once again called into question the form and meaning of the Lao state.

Projecting Laos

Lao hydropower gained global prominence surrounding the World Bank's involvement in the Nam Theun 2 hydropower project, the large tributary dam mentioned at the beginning of this chapter. The dam received considerable interest from investors in the early 1990s and was a critical component of countrywide economic planning by the United Nations Development Programme and the World Bank from as early as 1991. However, the Asian financial crisis called into question the dam's financing just as the transnational environmental politics of large dams was reaching a crescendo; the project was already described as "long-delayed" in 1997 (Woodrow 1998). By 2002, the Nam Theun 2 project emerged as a much larger dam, requiring more financing and a host of extensive ad hoc loans and conditionality agreements designed to fit the project within an overarching reform agenda. In this new arrangement, the World Bank and Asian Development Bank sought to coordinate as much as possible to present a synchronized set of multilateral development requirements: Laos was to be a model country for the demonstration of a revamped, sensitive, participatory neoliberalism, and dams would be the carrot motivating an extensive reform agenda.

Internationally, the dam was on stage as a test case in whether the World Bank could return to dam building after a hiatus of over a decade by answering to its environmentalist critics (Mallaby 2006; Cruz-Del Rosario 2014; Schaper 2007; Goldman 2005). The project was prominently featured in media such as *The Economist* and the *New York Times*, and as late as 2015, Robert Zoellick, former World Bank president, was publicly touting the dam as evidence of the bank's success in environmental governance matters. The issue was not that dam building had completely stopped during the 1990s—it had not—but whether, through an explicitly public and participatory process, the bank could set up a development package in which hydropower was part of an overarching reform agenda that showcased a vast array of new or reworked development tools.

A whole series of concatenated attempts to "environmentalize" the Lao state gained traction as the bank deliberated on its support for the dam project (Goldman 2005, 201). These included public expenditure management, which sought to establish transparency rules over government spending for external scrutiny by the IMF or other lending partners; a politicized debate around "policy-induced poverty," including a major anthropological study commissioned by the Asian Development Bank to show the damage caused by government policies (ADB 2001); and—especially relevant for hydropower—a vast expansion of the government's environmental regulatory capacity. The promise of sustainability attached to a US$1.4 billion project provided a major lever for these sorts of explicitly neoliberal reform efforts.

Yet it may be a mistake to assume that so much can be attributed to the program of neoliberal reform, and it is worthwhile to shift to an ethnographic perspective to provincialize the World Bank's brand of green governmentality. Sarinda Singh (2011) shows that access to foreign development aid, for government officials nominally attached to the conservation offices of the Nam Theun 2 project, forms a practice of patronage and allegiance and may result in pilfering of revenue from the sale of electricity. The question of the postsocialist Lao state is very much at stake, to wit: should the reforms be understood as genuine, or are they undermined by corruption? I concur with her treatment of the bank's green neoliberalism, yet I remain cautious about distinctions such as plan versus implementation, professional versus patronage relations, or improper use of funds. Moreover, much of this debate is too caught up in whether we can recognize the discourse of state functionaries as evidence of neoliberal subjectivation. Rather, neoliberal adjustments can be viewed from the perspective of emergent Lao postsocialist commitments to openness. The proliferation of discursive and practical spaces from Vientiane to the high plateaus of central Laos, essential to the ongoing project of "market transition," does not involve the subversion or redirection of formal plans so much as the constant repetition and blurring of a series of "amalgams and contradictory political juxtapositions" (Obarrio 2014, 33) that seem necessary for understanding experience but are difficult to sustain when scrutinized. Anthropologists can treat these experiences as part of the "backstage ethnography" of the Lao state (Petit 2002). Binaries such as formal/informal, plan/implementation, and discourse/practice are not analytical distinctions but practical evasions that enable the developmental state to function amid heterogeneous spheres of influence. By replicating across diverse spaces of development practice, they sustain confusion as a tactic of postcolonial survival.

Moreover, the apparent realism of anthropologists' attempts to document "actual practices" of the state has the tendency to replicate a certain desire for visibility at the center of the reform policies Goldman has documented. After all, the fear of opaque, informal state relations is at the very center of neoliberal discourse (Dwyer 2013), and the ethnographer's role is not to show what is "really happening" behind closed doors. Nonetheless, what Singh is responding to is essential for a situated appraisal of neoliberalism in action, for neoliberalism persists within an old program of visibility integral to modern forms of power (Crary 1999; Mitchell 1991; Mirzoeff 2011). For ethnographers and development practitioners alike, it is difficult to gauge the importance of the never-ending stream of development projects and multilateral policy pronouncements, many of which do not seem to have any real meaning. The ethnographic dimension of this problem is palpable. Many people in Vientiane's expatriate, English-speaking development circuit felt that the activity of the Lao state remained opaque and inaccessible, whereas multilateral policy and state planning pronouncements seemed simply to reproduce a manner of discourse that mimetically conformed to its own expectations. I am not trying to characterize expatriate development practitioners as a homogenous class—for many were ultimately highly committed to the legitimacy of the government—but rather I only want to show how neoliberalism manifests experientially in part by a desire to scrutinize the inner workings of the Lao state and that experience frequently entrains anthropologists as well. Reciprocally, the state seems preoccupied with looking like a state in the eyes of foreign donors. Seemingly only concerned with maintaining appearances, the putatively real state comes to appear ever more remote or inaccessible. This conundrum stems from the fact that the state is the constitutive construct, simultaneously target and blind spot, of neoliberalism itself.

This prevalent suspicion of the Lao state by Anglophone development experts took a range of forms and was not limited to institutions but could quickly shift to a characterization of Lao experts or bureaucrats or Lao culture. One development expert noted "a strong sense of the unspoken [among Lao experts], even in face-to-face interactions. In meetings, there are always knowing glances around the room—people reading each other's reactions, like everyone knows something you don't." Much to his dismay, one young United Nations staffer found he had developed a "neocolonial attitude" from working with Lao government officers on a daily basis. "Everything is political. Frustrating, it feels like sheer ineptitude. My [official government] counterpart argued that Laos needed to institute a policy because all the neighboring countries had done so and they would otherwise be embarrassed. They just want to save face. I couldn't believe the other Lao took that explanation as a matter of course." Foreign development practitioners' experiences across these instances were that they live in their own reality, in which policies and discourses form an echo chamber, while the actual practices of the government remain inscrutable.

At times, the structure of feeling was explicitly racialized. One long-time consultant, a British man who had spent decades in Africa, gave me a dramatized, fantastical performance of his opinion of officials. "When I sit down with ministry officials, I challenge their professional credentials. 'Take your tie off,' I tell them. 'Forget about your academic degrees. You're a crook. That's how I see you, and this is how we're going to work.' . . . Every educated Lao is 100 percent compromised [by nepotism and patronage connections]." He had an explicit theory in which villagers and private developers—united because they are both ostensibly legitimate economic producers—should be allied against the rentier state.

A World Bank officer insisted that the bank's Country Manager "has been lied to so many times by the ministries. It is easy enough to get them to agree to reforms and to sign a memorandum of understanding, but it is much more difficult to monitor compliance and ensure that specific agreements are not abused." A mid-level bank staffer, his job was to meet regularly with ministry officials to review progress of key reforms, providing them with periodic surveillance and accounting of their progress, as well as solutions to the problems they encountered. "Most of the legal procedures are in place, but now the government must learn to implement them in accordance with international standards. They have to learn to take ownership for running their country properly. By working with them on a more everyday level of implementation, I try to encourage them to carry through with their reform promises."

"It is a critical point of huge importance," he wrote in a report on the legitimate use of hydropower revenue, "that the profits are transferred to [government] accounts and cannot be accounted for. As a consequence, it remains unclear whether the revenues from [hydropower] actually go to poverty alleviation and development projects as promised or are used for other personal interests. This circumstance raises serious questions about the development rationale behind the project!" The pathos of this modality of development expertise is left to circle around a hyperbolic desire to establish a regime of visibility.

Such routine suspicion is a feature of the active politicization of development in the name of neoliberal reforms, including demands for multilateral sustainability programs. The constantly implied illegitimacy of the state manifests what Foucault (in Dreyfus and Rabinow 1983, 187) once called the "local cynicism of power," and the self-appointed international development community interpreted its role in explicitly political terms. Furthermore, conditional aid and multilateral surveillance put governments in the position of perpetually attempting to "look like a state" and, indeed, of simplifying their own state apparatus for the purpose of external oversight (Viktorin 2008; cf. Scott 1998). Finally, there is a hallucination effect in which neoliberalism's "institutions, knowledge practices and artifacts thereof . . . internally generate their own reality" (Riles 2000, 3) to the point of replicating colonial patterns of misrecognition. Rather than a thorough problematization and restructuring of the state under a program of neoliberal governmentality, the would-be neoliberal state rather turns into a kind of mirage in which the native state is constantly imagined to be dark, monolithic, and hidden from view by the very projection of neoliberal reform imported by approved World Bank consultants.

Put differently, the proliferation of so many spaces and contexts of enunciation and negotiation has been part of governing the opening of Lao rivers to transnational investment and development assistance. Lao hydropower has come with a barrage of contestation, criticism, and bad advice. For a country that has been systematically manipulated by foreign powers, jealous of its hard-won independence, and proud of its consistent peacefulness, a crucial task is the tricky job of governing the unruly freedoms of those who participate in the country's collective improvement. Understood as a governmental problematic of managing the powerful asymmetries of foreign engagement, Lao postsocialism is as much about maximizing openness and growth as it concerns governing the powers of freedom expressed by transnational programs of investment and rule. This raises questions about the affirmation of hydropower from the vantage of Lao national expertise.

Affirmations of Expertise

The global environmentalism projected onto Laos served to multiply the diverse points of entry and technical forms through which Lao rivers took center stage in transnational development projects. I have already described the concessionary basis as crucial to governing transnational programs of investment and rule, but it remains to be seen how that affirmation of riparian development and political openness manifested among an emerging urban technical or intellectual class. A remarkable feature of Lao developmentalism during the postsocialist period has been the rise of an expanding class of young professionals working in government, international NGOs, or otherwise engaged in and cognizant of changing Lao political openness. This group, usually distant from real positions of influence but nonetheless with a stake in the workings of government, has increasingly articulated a set of normative expectations and affective investments vis-à-vis expertise, often times reimagining the Lao nation with an internationalist orientation while demonstrating the value of hydropower through their own professional work. They are able to think and work actively on their country's social, environmental, and economic relationships in ways unavailable as late as the mid-1990s and otherwise off-limits in the tightly controlled single-party state (for example, through public organizing). If it cannot be said that environmental norms or discourses of sustainability define the functioning of the government or the rationality of state bureaucrats, there is no question, nonetheless, that a powerful affirmation of expertise was increasingly integral to the work of government broadly construed.[5] Two features of the powers of expertise concern what might be called a *politics of listening* and a *practice of conforming* to international standards, both of which prominently feature the unstable relations with foreign organizations and expertise discussed in the prior section.

Politics of Listening

At the outset of this chapter, I discussed a situation in which a Lao filmmaker affirmed his reasons for not taking an explicitly critical approach to the government's handling of the environmental damage along the Xe Bang Fai River: his point was that, if the context is known, the criticism will be apparent without needing to embarrass the parties who have failed in their responsibility to protect Lao citizens. Perhaps that seems inadequate from the vantage of contestatory politics of the liberal democratic West. Yet it is evident that transnational activism depends on a politics of confrontation and voicing of opinions, whereas Vientiane's emergent technical public, in its relation to the party-state, places the emphasis on listening rather than speaking.

Openness was a fraught and dangerous project partly for the challenge of hearing expertise, as here voiced by Sommai, an environmental technical staff whose job was to implement the environmental laws put in place by the Asian Development Bank. "Don't be surprised that people in the government don't take all foreign expertise recommendations at 100 percent. Instead the government has to analyze these reports with their own strategy of what the people needs and what the government should give [them]." Here is a very similar comment: "The government takes all of the recommendations from all the different reports and reads through them all and selects what to adopt and what to not adopt. You have all of these reports that go in, if you can understand what comes out then you might have understood the Lao mentality."

Sommai was a technician in his early forties in the hydropower office of Electricité du Lao, the state utility, originally trained in Thailand, and he agreed to meet with me early on in my research to explain what that process looked like. In a rehearsal that would become routine, he wanted to understand exactly who I was connected to and what other Lao I had spoken with before our conversation started—and therefore who might be likely to hear his words later, in whatever garbled form I might put them. Sommai had strong words to describe the imposition of foreign environmental expertise on Lao hydropower, for a project that at that time was still three years from beginning construction: "We need greater flexibility in bank support. For NT2 the World Bank has requested NT2 to do so many different studies—first water quality, the impact of relocation, then they asked for more—biodiversity, maybe about ten different studies all associated with just one project. They weren't ever really happy. The project has been underway almost 10 years."

Environmental demands were experienced as a question of moving the goal posts of development and ignoring obvious economic needs. Invoking IMF structural adjustment reforms of the 1980s, a deputy minister's response to *farang* (white foreigner or Westerner) criticisms of government plans, here in an English-language public email forum, captured the chaotic experience of constantly changing foreign expertise: "NT2 was the biggest challenge but it was finally considered as a 'model way' of building dams. So this means that Laos is up to the task and never, in the history of Laos, have I heard that we have 'bullied' our neighbors. Yes indeed *farangs* have lots of opinions. . . . In the late 80s we were asked to 'privatize' all of state-owned enterprises and we were told that by doing this we would be very rich within only a few years. But this never happened and I never saw those guys coming back . . . boy do I wish to see those guys here again."[6]

In addition to being an explicit criticism of the model environmental project against the backdrop of IMF structural adjustment, this commentary is also a rare public display of an important modality of governing expertise, namely the rigorous effort to control public criticism. The socialist understanding of the role of intellectuals in society required "solidarity" with "the worker-peasant alliance . . . under the leadership of the Party" (Boupha 2002, 11). This understanding of expertise does not seem to have changed very much. However, the problems are inherently more complex, involving an exponential increase of foreign expertise, the intimacy of its engagement with state practices, and the expansion of an urban Lao technical class.

The deputy minister, uncharacteristically open about his opinions, took on an active role in admonishing outspoken younger critics of policy and practice. During the early 2000s, it was very rare to hear explicit criticisms on government policy on environmental grounds by Lao citizens, but by the height of the dam-building boom circa 2010, public criticism, in this case from a junior technical worker with his own clear political opinions, was possible: "I could say that we could be, in one way, an informal secretariat unit for our government of Laos since we have some technical experiences in the field working with the poor remote lowland and upland people rather than those office staff whose thoughts and point of views are always the top-down, meaning they do not even care whether Environmental Impact Assessment or Social Impact Assessment of the dam are well studied and prepared."[7]

His intervention was quickly and publicly challenged by the deputy minister, but the exchange is indicative of a manner of state listening that is remarkably attentive to the details of what people say. The idea of organizing a citizen-led expert "secretariat," seemingly innocuous from Euro-American standards, undoubtedly was very carefully scrutinized and managed by the government. During my research, I came to understand that Lao staff working for international organizations, many of whom are in their thirties or forties, were annually hosted by the government for an event in which they were exhorted by government officials not to betray their country's secrets to the foreigners they worked with. Singh 2012 documents the systemic stifling of criticism in which villagers expressing doubts about "voluntary" resettlement were literally shouted down, in the intimate setting of the village, for violating party orthodoxy in the presence of a foreigner. The most striking case of this kind of silencing, which can be easily interpreted as a tactic for governing foreign-influenced expertise, is the disappearance of human rights advocate Sombath Somphone and expulsion from the country of the Swiss director of the Helvetas charity for their advocacy of land rights (Baird 2014).

Such evident repression therefore shows the rise of a nonelite urban technical class that increasingly has a stake in how the country is run, the government's sensitivity to criticism, and an unexpected pragmatics of listening often simplistically misinterpreted as management of appearances. These are commentary on the role of independent expertise and the emergence of a public where political viability is held to rest on national unity (Michaud 2013) and balance (Ivarsson, Svensson, and Tønnesson 1995). Here again is the deputy minister, whose unexpected candidness nonetheless represents a completely orthodox perspective: "There are some independent NGOs, the so-called 'environmental activists' [*sic*] which will never agree on dam building. . . . So based on my own experience working in watersheds and in preparing for NT2, I have learned one important thing: don't listen to these guys because they don't want any good for the country. This is my real and true past 30 years of experience."[8]

Echoing Sommai, who said that the government has learned to parse carefully the advice it receives, the deputy minister's advice is quite literally to *not listen*. As Candea (2014) might put it, "this, too, is politics, albeit not 'our' version of it." Whereas Euro-American understandings of democracy and public life are predicated on ideals such as free speech and public assembly, which at first glance seem almost nonexistent in Laos, I would argue that government officials and many of the emerging technical class are averse to transnational environmental politics because it is articulated so loudly that it cannot be heard. Put differently, if political practice seems muted and indirect, it is because it depends far less on having a particular opinion and far more on a politics of listening and discretion in the service of national unity.

Practices of Conforming

Openness has given this emerging technical class a strong internationalist flair, even while they push the limits of how they can participate in affairs of state. Again and again the question of international standards came up in the context of national shame or embarrassment. This is how Sommai described the process of developing environmental standards through the regulatory reform of the state environmental agency: "If the Lao government doesn't have its own standards, it has to refer to another standard that is perhaps disputed. In many respects they are learning through doing, and sometimes they hesitate to apply the law strictly because if something goes wrong they will be responsible and thus embarrassed."

International standards reference an "ontological subjectivation," perhaps visible only in outline, of *conforming* as a process of learning and growing, of subjecting oneself to power relations—of the possibility of composing an achievement. Dams are hard to build properly and to do so is a clear sign of national independence.[9] The subjectivation is ontological insofar as it involves learning how to accommodate the requirements of building dams. It is demanding to conform to international standards, yet it is through those standards that rivers maintain their seductive possibilities. This ontological subjectivation is essential to understanding the forms of skill involved in working with things.

One engineer for the national electricity utility described to me his recent certification by the International Organization for Standardization as a kind of guarantee or assurance of his status as an engineer and source of professional pride. Alone the certification seemed commonplace, but the question of intimacy, correct execution, and deliberate conformity came quickly to the fore. The electrical engineer shared with me photos of his young son, six or seven years old at the time, carefully assembling a model truck from a kit including a sheet of detailed instructions. He had placed them in a photo album titled "Son of the Engineer," and they included a series of several images of the young boy diligently following the step-by-step instructions, finally showing the assembled model truck alongside the sheet of instructions for assembly. He had written a caption in English that makes clear the ethic I am trying to identify: "You are the son of the engineer. You must understand the plan. Read the plan and assemble the model. Do it yourself. Good. I'm proud of you, Son." In this familial intimacy I hear a commitment to development expertise, to standards and planning, and to independence. But this intimacy goes further and demonstrates how transnational openness and developmental modernity are saturated with personal commitments that mark the possibility of composing an achievement—kinship, gender, project, and nation as targets for investment and work.

Cast within a history of marginalization, achievement defines a temporality of self-formation. Vatthana was a thirty-two-year-old human resources manager for a major international development NGO in Laos that had done consulting work on the hydropower dam I was studying. She conveyed a personal sense of the "international standards" that accompanied her work in development—albeit in a nontechnical sense. "My husband worries that I'm not 100 percent Lao. It's like I'm half international, but I'm not pretending, it's automatic. It also comes from going to school in Australia. Even if you're rich in Laos you may be educated, but if you never experience life in foreign countries your style will not be up to standard." She was asked to design and implement performance evaluations for NGO employees, hardly a popular proposition, which met great resistance from her coworkers. As she later told me, "When I started it was the most difficult thing I'd done in my life," she said. "I thought I would have to leave [the NGO], it was so stressful. I didn't think I could do it. . . . But if I was to leave, how would I show my face when I tried to find a job?" Just as with Sommai's comment about applying the law incorrectly leading to embarrassment and to the experience of pride when the engineer's son successfully executed the plan, Vatthana directly references an experience of shame or embarrassment at the thought of failure.

She provocatively described the challenges she faced conforming to the standards she set for herself.

> Many Lao people find it too difficult to adapt themselves—to change yourself even a little. If you're satisfied with your spirit and soul you can change yourself. You must understand the situation, you must have mature thinking, understand what you're doing, understand your goal. Personal goal, personal objective, enthusiasm, personal thinking, personal motivation. Give yourself motivation. Sometimes when I feel really down, I say a word out loud to myself. Some people just think it to themselves.
>
> People ask me, Vatthana, why are you so calm? People see me as calm—but I think I'm not . . . if I'm not happy it's like I'm going to explode. Like I'm cutting off part of my body for team work.

The affirmation of Lao hydropower is not so far removed from questions of international aspiration, personal motivation, and embodied experience. Vatthana's description of the rigors of transformation, I think, is not incidental to the overarching predicament of Lao hydropower, which perhaps felt somewhat like a way to cut off the country from its past. Sommai argued that backward villagers needed to be educated and invited into development and could not be left to develop only if they wanted to. "People in the Biodiversity Conservation Area, they have a very low living standard, which is under the poverty line and primitive as 500 years ago, how can it be that they don't want development, as claimed by the consultants? . . . Consultants sometimes seem to want a sort of human zoo, full of primitives where *farang* [Westerners] can come and look."

Similarly, the physicality of self-formation (or deformation) manifested in gestures and nonverbal cues demanded of me the challenge of listening in unfamiliar ways. Sommai's disposition as a technical staffer was perhaps best indicated by his refusal of the breadth of some of my questions. He rather wanted me to ask much more specific, precise questions as his gesture indicated a limited domain or scope for what is appropriate. The question of international standards was continually at stake. As the passage continues, let me suggest that he uses environmental regulations, conforms to them, and indeed holds his body in relation to them as an instrument integral to his task as regulator. When I asked how the ministry manages foreign hydropower developers, he described the Environmental Impacts Assessment process: "Your question is very broad but the answer I will give is very specific [holds forefingers apart by about 50 cm and then brings them close together while speaking]. The EIA regulation is quite new, and when an EIA is submitted it can be judged with reference to the regulation. [The government] might request that paragraphs of the EIA be translated into Lao so that they can better understand what the document proposes, such as a section dealing with involuntary resettlement. The regulation allows them to require this of the developer."

In this passage, Sommai discusses learning how to use the environmental law, which has only been in place for a couple years. He worked not only to make his practice conform to the requirements of the law but also to make his country conform to international standards. The riverine imagination of Lao national development takes form in the work of international organizations, technical standards, foreign investment, and the multiplicity of institutional structures through which these function. These in turn are undergirded not only by the infrastructures of electricity but also the activity of their risky projects. The activity or process defines an obligation of worth oriented toward certain outcomes rather than others. That is, the value of becoming is inscribed in the process of an attempted technical achievement.

This confluence of the intimacy of expertise and the promise of achievement helps demonstrate why hydropower remains so central to an affirmation of Lao development. "Hydropower," Sommai said, "is expensive to plan and implement, because you have to prepare for it very well in all aspects. But it has a much wider circle of importance than smaller projects [made circle with hands in air]. Dams bring in lots of benefits—infrastructure, roads, hospitals, schools."

Sommai's gesture marks hydropower's circle of influence. Encapsulating the logic of the foreign concession within an affirmation of national development, expertise, and international standards, Sommai's gesture grounds the promise of hydropower within a nationally defined circle of influence, rather than those spheres of power to which his country has been subjected throughout much of its modern history.

I have used the image of the circle of influence to capture the diversity of political spheres that have come to bear on Laos's rivers through the course of a long, ongoing saga of political emergence. Between the French colonial and American imperial regimes, the Lao People's Revolutionary Party could exist only in the powerful ambit of Vietnamese and Soviet support and protection. Liberalization promised a path to economic development, and therefore financial autonomy, only by subjecting the government to the strictures of IMF-led adjustment programs and intensive multilateral programs of normalization—while opening the country's resources to partition and commodification by foreign capital. In contrast to weak and ineffectual development aid schemes, the benefits of hydropower are tangible and long-term, even if they come with evident costs. Large dams enable the imagination of a powerful circle of influence that emanates from Vientiane to better define the country's postsocialist independence.

Indeed, this strategy of maximizing relations with foreign powers has a long and provocative history integral to the practice of precolonial, sovereign kingship. While it is impossible to describe the connection in terms of a straightforward cultural continuity, there are clear echoes of classical Southeast Asian political forms. Classical Tai polities functioned through concentric and uneven circles of power that were territorially extensive but defined by control over populations rather than definitive territorial boundaries. Described by Tambiah as the "galactic polity," this political form involved a metaphysics of power associated with "big man" charismatic rule (in current Lao, *phu nyai*) and the mandala form of organizing political space in terms of multiple sovereign spheres (Tambiah 1976; Wolters 1999).[10] There is no *simple*, direct cultural or historical connection to contemporary Lao politics, and unlike certain observers (Stuart-Fox 2005) I find it impossible to deduce from that history a present-day culture of corruption based on clientele relationships.[11] Nonetheless, if the strategy of maximizing relations with diverse foreign powers bears analogy with precolonial Lao political forms, it demonstrates the extent to which hydropower is affirmed as essential to the viability of the country's future.

Conclusion

A powerful confluence of prerogatives has come to bear on Laos's rivers, showing the anthropogenic force of rivers as they concatenate these diverse political projects. Dozens of dams are now proliferating along these rivers. But even long before they are built and operational, the process of assembling those dams has produced a many-faced sociotechnical drama that constantly affirms hydropower development and bends the social world around its aspirations and requirements.

Essential to Laos's program of economic diversification has been a distinctive political strategy of drawing out the complex commitments of powerful outsiders. One purpose of this chapter is to show how Lao national expertise has a limited and specific role in the sustainability enclaves of transnational resource investment. Due to circumstances surrounding transnational environmental activism, Anglophone managers and activists came to have a predominant role in governing the ecological effects of the dam that concerns the remainder of this ethnography. Attempts at state reform and the private indirect governance that encircle the sustainability enclaves recall certain features of Laos's historical marginalization. Yet there is an essential difference. The current configuration hinges on a strategy of economic and social openness, an engagement with all sorts of transnational actors, escalation in the content of government and its commitment to knowledge, and proliferation of many discrete spaces of transnational investment and rule. It is impossible to ascribe Laos's riparian natures solely to a totalizing mechanism of extractive capital or an imperial program emanating from the bureaus of multilateral development institutions, whether separately or in combination. Although important, neither of these explanations takes seriously Lao political rationality, its affirmations of development, or the seductive potential of the country's rivers. Moreover, if one can say that Lao rivers have become neoliberalized, the inverse is also true. Due to its intensive work to incorporate rivers into its political strategy, Laos's neoliberal postsocialism is distinctly riparian.

In this chapter I have shown how Lao hydropower, through a process of neoliberal market transition, came to be a crucial site of intensive projects of transnational investment and rule. The transformation of Laos's rivers underway by the 2010s was preceded by two decades of experimentation on contractual risk-sharing agreements, spatial articulations of foreign investment, legal and regulatory reforms of the Lao state, and attempts to secure a particular brand of multilateral environmentalism. It also formed an essential component of the strategy of openness or "renovation" through which the socialist project was reconfigured within a rapidly shifting geopolitical situation. These diverse investments in Lao rivers called into question the contemporary form of the Lao state in contradictory ways. Whereas the neoliberal project was oriented toward a project of enrolling the Lao state within a standardized template of hyperlegibility, the government of Laos put into place a mode of concessionary rule that essentially delegated certain kinds of environmental government to the sporadic and highly variable interests of diverse foreign actors. At first glance Laos appears to be a small, remote country of little consequence, but, because of a long-term historical process of marginalization and dependency, its rivers have come to be the site of intensive, cosmopolitan experimentation.

The affirmation of large dams within the government's development strategy invokes a diagnosis of the country's complex postsocialist predicament. With a hydrological imagination forged in the crucible of American military domination, large dams stake their claim on a plausible political future and an opportunistic attempt to establish the country's viability. Like the power relations to which Vatthana expressively commits herself, large dams forge an unstable but novel assemblage of political possibilities, require the conditional acceptance of certain power relations rather than others, and make powerful claims on transnational political economy. Moreover, in the midst of a process of wholesale transformation, Lao rivers place stringent demands on engineering, financial, and environmental knowledge, and these types of knowledge in turn play a decisive role in configuring the sociality of hydropower. The affirmation of large dams imposes direct, historically specific, and nondeterministic obligations on the practices and reason of those who come to participate in their composition. There is an obligation *to* rivers that comes with their imaginative, opportunistic appropriation in the service of human designs. That obligation takes contingent form in everything from risk-sharing agreements to the habitus implicated in environmental standards.

The production of ecological uncertainty that characterizes late industrial environments, as well as the practices of technical entrepreneurialism I have claimed are indicative of Lao hydropower, appear *en force* in the coming chapters, for they are elements of a privatized micropolitics that remains largely invisible in governmental dilemmas of the central state. But those dilemmas also always anticipate a politics of uncertainty in which opportunity and threat are constantly in play. The regimes of authoritative knowledge that characterized hydrological planning in the 1960s and 1970s have mutated into something complex and frequently unrecognizable. The modernist dream of large dams was to establish a lasting basis for the foundational security of the modern sovereign state. By the late 1990s, the pluralism of Lao rivers was predicated on transitory amalgams rather than comprehensive plans and on the multiplicity of reason rather than its putative singularity. Even the World Bank's approach, which looks comprehensive and singular at first glance, is oriented in practice toward managing images and proliferating contradictory discourses, each with their own distinct constituency and history. Whereas the planning exercises for the Mekong basin in the 1960s were the work of a relatively small cadre of technical experts who mapped the largest, most ambitious possibilities for the region, sustainable hydropower incorporates a cacophony of voices, explicitly invites critics into the process, and multiplies the sites of investment. Those are anthropogenic productions taking emergent, historically specific form. The World Bank's approach also produces representations of dams, their publics, and their natures at orders of magnitude greater quantity and level of detail than the expertise of the 1960s and 1970s, and yet that expertise frequently fails to consolidate an authoritative representation. To that extent, sustainable hydropower requires a production of uncertainty that is not ensured by the authoritative confidence of planning or the verisimilitude of expert representations, but rather is predicated on anticipation and promise.

Anthropogenic rivers are profoundly social long before they are transformed into late industrial environments. Dams' effects precede them, for they operate within a temporality of anticipation. The vast majority of the activity discussed in this chapter occurs in the imagination of hydropower dams and their promises, not in the wake of their construction and operation. By describing specific forms of practice and reason organized around these rivers, we see that their heterogeneous combinations persist in the temporality of promise and threat—of opportunity, risk, and danger—in which protracted debates surge over who has the right to make such promises. These power relations can be described partly in terms of contestation and struggle for hegemony, but they are also already the formative effects of dams yet to be built. The forms of the human that proliferate in relation to Laos's rivers forge a relationship to rivers of the future.

Neoliberalism's natures depend not only on a discursive rationality that naturalizes the world in its own image, but also on the seductive promises of ecological events, such as the building of dams, in which the powers of anthropogenic rivers are harnessed in ways that cannot be controlled or even anticipated with much specificity. This production of uncertainty, and the privatized mode of practice I am calling technical entrepreneurialism, will become powerfully apparent in the chapters to come. For the circles of influence that encompass Laos's rivers, the production of uncertainty includes the anticipation of vast opportunities for resource investment and national development.

Rivers are not determining. They do not impose limits or dictate requirements, or rather those requirements are always articulated in terms of historically specific repertoires of knowledge and imagination. People conform *themselves* to the demands of rivers through the long-term development of multifaceted infrastructures for building dams, replete with practices of imaginative anticipation and technical acuity. That imagination is a multiplicity: there is not any one way to build a dam, nor only one way to critique dams. Conversely, the ecology of large dams is enacted through the staging of an ecological event that effectively comes to permeate and reorganize a whole series of relations. There is no single cultural meaning or system of power to which large dams ultimately may be reduced; nor are they simply vessels empty of meaning into which people pour significance. Rather, they accrete all sorts of diverse, partial affirmations and negations, including those that seek to hold onto the tenuous possibility of maintaining a way of living in their wake. Those partial connections are already part of hydropower's riparian ecology. Through historically specific forms of strategic opportunism, powerful new vitalities are wagered on the definite, real but nondeterministic ecologies of rivers. When rivers talk, they whisper dreams of power.

Interlude

What Is a Dam?

At the center of the Theun-Hinboun Reservoir is the confluence of two tributaries, the Nam Gnouang and Nam Theun. I arrived here for the first time, roughly in the middle of my research, by boat—traveling downstream from the headwaters of the Gnouang River. This was very nearly the first time I had even seen the reservoir. I also had never seen the dam. The Gnouang was heavy with silt from the powerful monsoon season, and it slowly mixed with the darker water from the Theun as the swollen mass of water moved toward the dam downstream. During the monsoon, the water cascades over the crest of the dam, far more than can be used by the power station. When the dry season comes, the downstream release is clenched to a trickle, five cubic meters per second, while the rest is piped below a mountain ridge to the power station and into another river, the Hai-Hinboun, and into another space, another time. We think we know what a river is—a body of water, a powerful current, a moment of gravitational flow in a biospheric cycle—but a late industrial river is difficult to know.

The Theun-Hinboun Dam was originally called Nam Theun ½. Naming conventions come from Cold War–era river basin planning. In the high modernist period, hydropower planning was a heroic effort to maximize the comprehensive utility of whole river basins. Teams of engineers—in this part of the world, American engineers from the Bureau of Land Reclamation and the Mekong Committee—created rough plans for the most rational organization of the maximum number of dams within a river basin. Many dams are mutually exclusive; it is best to build those that together lead to the maximum output at best cost for the basin as a whole. That was a time before privatized investment and joint ventures, which tend instead to fracture the river systems. The Theun-Hinboun was not part of that original planning. It was named Nam Theun ½ because the site was identified, in the early 1990s when planning started and things got underway, between Nam Theun 1 and Nam Theun 2—two massive dams that did not yet exist. And then there was Nam Theun 3, located further upstream on the Gnouang River, which is a tributary to the Theun River and therefore part of the river basin. These infrastructural landscapes have histories and therefore poetics (Larkin 2013). Unpacking these infrastructural histories provides the context for the ethnographic rendering of materialist concerns of people such as farmers, engineers, and project managers.

For the developers, Theun-Hinboun was a serendipitous coup. Absent from the comprehensive planning, the site was identified by happenstance when engineers were flying back by helicopter from a site visit from the Nam Theun 2 site (this story may be apocryphal). The site itself is monumental. From a high-elevation alluvial plateau, the massive river cuts through a narrow notch between two ridges—the perfect location for a dam. The only problem was that the dam would need to be small or else it would have to be huge. For private investors, that was fine. Build a small dam quickly with little or no resettlement, earn a large return on investment, and avoid the fuss. Leave the big projects, like the much bigger, upstream Nam Theun 2, for the World Bank and the activists to squabble over. When the Theun-Hinboun was built, Nam Theun 2 was very much in process and everyone expected it would be well underway probably by the year 2000. Nam Theun 2 takes water out of the river and, like Theun-Hinboun, routes it to a totally different river. "They're stealing our water!" I heard a dozen times from Theun-Hinboun Power Company managers. The government later gave the

Theun-Hinboun Power Company a consolation prize—the undeveloped Nam Theun 3 project, upstream along the Gnouang River, and a lot of my research took place under the auspices of helping environmental scientists for that expansion project.

We had been working along the upper reaches of the Gnouang River. I was embedded with a team of environmental scientists who were conducting surveys for Nam Theun 3, which was to be an extension of the Theun-Hinboun hydropower project. The upstream reservoir would store perhaps five billion cubic meters of water for use in the dry season when the Theun-Hinboun Reservoir did not have any water to spare. We established a base camp at Hat Sai Kham, a small village at the end of a long dirt road and three days' journey hiking and by boat from the upper watershed, marked by the border with Vietnam. Later we would find that there was a military road accessing the area. It was not on our maps and was strictly off-limits.

James was the head of the team, an aging British zoologist who had lived a storied life tracking animals in eastern Africa before setting up an environmental consulting enterprise in Laos with his sons. He was a utopian of sorts, for he believed that hydropower could be harnessed for large-scale social engineering, if only one could remake the whole of relations in which environment and society are enmeshed. This utopianism took many forms. One of my favorites was a massive Excel spreadsheet in which they attempted to pin an economic value onto each village-level environmental "good," that is, each kind of resource collected from the forest, each fish species, and so on. Everything in theory has an economic value, his son told me—the challenge is to find it. They wanted to use this to create realistic estimates of the actual damages caused by dams. We worked with one of his sons, several porters, and three Lao research assistants, Tong, Phat, and Singphet. From Hat Sai Kham, we traveled upstream by boat, initially on "reconnaissance" missions, conducting village surveys and attempting to get an overview of the issues for villages that might be affected by the future project. Of all the environmental consultants I worked with, with the exception of certain fish specialists, James was the only one to insist that he was a scientist whose task was to produce objective facts.

Upstream water flow is determined by downstream conditions, I learned (I am still unsure whether I believe it). We were above the elevation of the reservoir, but still the water was sluggish. James and I traipsed through bush around an out-of-the-way trail leading between two villages. A karst massif towered behind us—sheer limestone cliffs, intricately eroded with jagged edges and caverns that twist and disappear. Karst is impenetrable to people, and new species are routinely identified near it because there is so much potential for isolated, protected niches. James was trying to figure out whether these marginal lands are used productively or not. Some plants seemingly were cultivated or at least favored by human intervention, but neither of us recognized them and the team plant specialist was not with us. Truth be told, James was looking for marijuana. The night before, the village chief slyly told us that the village used to grow marijuana in a remote area no one could access. When he learned I was studying in California, the chief excitedly told me his niece lived in Fresno. He found a small notebook and wrote down for me his niece's address and phone number. She serviced vending machines, apparently, and I was impressed that his niece has had the patience to communicate such an odd idea as a vending machine across continents to her uncle.

We were not in a remote area in which someone might surreptitiously grow illegal drugs, but there we were, traipsing around. Through the brush, we broke onto the river. Late in the afternoon, the red sun was haloed by dramatic clouds. James waxed eloquent about the natural beauty of the landscape. He speculated on ecotourism.

The environmental scientists were contracted to gather socioenvironmental data and begin analyzing prospects for managing impacts, but a major, ad hoc task for the team was to determine the volume of flow in the river. Water volume is a matter of hydrological risk, since the amount of water is directly related to amount of electricity that can be generated. This question has nothing to do with sustainability or mitigating environmental harm, only with whether the dam will generate electricity. No water, no power; idle turbines are lost revenue. James hired young village men to measure the river's water level using handheld barometers, twice a day, at designated points along the river (figure I2.1). The barometers must be calibrated with each other, and then variation in the air pressure due to weather changes must be corrected for. For this there was a reference barometer, set against a river gauge, that showed consistent but relative values. Will these young men really go down to the river twice every day to take the measurements? Who will calculate the data corrections? Is it reasonable to assume that weather patterns will be homogenous over the area? The enterprise seemed far-fetched, but it foregrounds the role of tenuous, uncertain number practices

Figure I2.1. Training a village youth to take twice-daily barometric measurements, calibrated against a river gauge. Photo by author.

to the engineering project—and the ways certain people contort their bodies to afford rivers their wishes.

The number work is shot through with investment risk. We made it back to the operators' camp based at the existing powerhouse, where James's team used a staff apartment for an office. A plastic storage bin had been turned into a makeshift specimen box, with a few inches of water holding a fish that needed to be identified. Stacks of hard drives lined the table. We met with Raymond, the project manager and old friend of James, over a meal in the canteen. Raymond was a sometimes boisterous French Canadian engineer hired from Sogreah to oversee the many facets of the project—the same engineer who had worked on the Martyrs' Monument in Algeria. He had written a technical evaluation of the design tender proposals. Recently convicted in a high-profile corruption case in Lesotho, Acres International scored the lowest for the quality of proposal. It was also the cheapest. Once again, numbers about water played the critical role.

The company hired Acres anyway. This was important for two reasons. First, the Acres engineer on site was pushing the company to build a different dam—a completely different design, in a different location—than the project under consideration. Second, for the project that the company had planned to build, it turns out that they had no idea what volume of water the reservoir would actually hold: the plans put the number at five billion cubic meters, which was clearly very wrong.

A cubic meter of water is roughly a metric ton (1000 kilograms), adjusted for temperature.

Five billion cubic meters is a lot of water. The existing reservoir was estimated at fifteen to twenty million cubic meters, that is, almost three orders of magnitude smaller. But why were the plans wrong?

Snowy Mountain Development Corporation seems to have created the five billion cubic meters figure, probably in 1991 or 1992. In 1991, Laos had just formally liberalized. Engineering companies swooped in, only some of them serious about actually building dams. Rather, it was a speculative market of staking claim to specific projects and winning concessions, which could then be flipped to actual investors. To win the concession, the company must complete the initial design work for the dam, which is what Snowy Mountain Development Corporation did. Using conflicting sets of topographical maps, they selected the set of maps that gave the largest reservoir volume. If another investor was going to buy the project from them, then five billion cubic meters of storage would help set the price.

Snowy Mountain Development Corporation never sold the project, and the concession expired after ten years. As prospects for Nam Theun 2 ramped up, the government gave it to Theun-Hinboun to develop, and Nam Theun 3 became the Theun-Hinboun Expansion Project.

Back at the canteen table, with James and Raymond, there was intense but quiet speculation about the motivations of the Acres engineer, who wanted an even bigger project, Raymond said. I interacted very little with the man. He was only here for a week. Snowy Mountain played with the contradiction between two sets of maps, and Acres played with the fact that no one knew what would be the volume of the reservoir. Without the data to be plugged into a computer model, it was difficult to estimate the reservoir volume. A rough calculation showed the actual volume might be 1.5–2.5 billion cubic meters, based on the other maps. At any rate, if one map is wrong, why assume the other is right? High-precision GPS elevation gauges were being ordered. In the meantime, the practical consequence was that the possible location of the dam has been totally opened up. The question thus became, of all the possible locations, which will give the best combination of cost, risk, and return? James was tempted by this location just where we had been rooting around, looking for marijuana. The karst massif on the south side of the river outlined what could be a tremendous reservoir; another ridge defined the north shore. Looking at the map, the site was exhilarating. Find the contour line, trace the reservoir outline—a cartographic imagination. Whereas the other projects trace long, narrow reservoirs that snake up the mountainous riparian zones, this project would result in a vast, expansive lake. Raymond argued that it is impossible. The geological risk would be far too great with the karst. There is no way to know if there is some major cavern system that will drain the reservoir directly into an aquifer. There is no way to tell where that water might come out, he says. Nonetheless, that option later became one of seven listed in a formal environmental cost and risk-assessment matrix.

In the subsequent years, as I left to write my dissertation, Raymond, in his early fifties—and an easy-going supporter of my research—died of a heart attack. James entered a pitched battle with the project developers, for he wanted them to fund a massive irrigation system below the dam in order to convert the whole area to irrigated dry season rice—another of his grand social engineering ideas. When the developers pushed back, he started leaking documents to International Rivers. Discredited, his contract was revoked, and from that point on I had no particular inside access to the field site.

Several years later, I received news that James was kidnapped on assignment in Somalia. His company with his sons had relocated to Kazakhstan following on intensified conflict with hydropower developers and the Ministry of Energy. From his son, I received periodic appeals to help put pressure on the British High Commission to locate him, plus news of occasional (but unconvincing) demands for ransom. However, he was also shot in the leg during the kidnapping. Already in his seventies, hopes had faded for whether he was even still alive.

To my surprise, the dam was built in the location James had argued for that day in the canteen—the biggest, riskiest option. I do not know how geological risk from the karst was addressed. As it stands, hydropower dams are built on a bevy of "far-fetched facts" (Rottenburg 2009), but such shoddy numbers are not merely approximated—they are opportunistically affirmed and promoted. It is no surprise that many projects suffer cost overruns, technological failures, or financial ruin for their public backers. By the same token, the technological apparatus of the dam is not opposed to ecology; it is an unexpected iteration of the river that affirms a certain, small number of ecological parameters at the expense of others. It is a singular and improbable intensification of that river.

Chapter 2

Vulnerable at Every Joint

It was not wholly accidental that, in the mid-1990s, the nascent Lao hydropower industry was so unprepared for the activism of IRN. In 1998, the year the Theun-Hinboun Dam began commercial operation and the year IRN's campaign against the dam took form, transnational antidam activism had marked the rhetoric and planning of megaproject development for over a decade (Goldman 2005; Baviskar 2005). By that time it was well known how the World Bank had been shut out of the Narmada Valley Project in India due to environmental and social justice activism and was subsequently forced to cancel a major project in Nepal. The World Commission on Dams, described as "arguably one of the most innovative and even unprecedented international governance experiments in the area of sustainable development of the late 1990s," was already underway (Khagram 2005, 146).[1] Popular and environmental protest in Thailand around the Pak Mun River development forced the Thai government to idle temporarily the power generators under court order—a massively expensive compromise—and thereby provided a major financial impetus for foreign investment in Lao dams, since financing dams in Thailand subsequently was impossible well into the 2000s (Missingham 2003). Even during the construction of Theun-Hinboun, their environmental plans were challenged by a Scandinavian NGO. The company made some important design changes, but to developers activists never seemed like a serious threat. Dominated by multinational engineering firms and national politicians eager to tap into new lines of foreign investment, the industry in Laos was undeterred by so many megaprojects interrupted. Any of these events should have proved the high financial stakes. Yet on paper, Lao dams looked like they would cause minimal damage and, with no organized civil society, the government's part of the bargain included managing any social and political fallout. What developers did not count on was the range of tactics at activists' disposal or the sensitivity of multilaterals to public disapproval.

In this chapter, I shift focus from questions of national development for a political elite and its relatively affluent technical class to the narrow confines of a single hydropower company, the Theun-Hinboun Power Company (THPC), and its protracted negotiations with the noted activist group IRN.[2] In this narrower setting, it is possible to assess the events through which Lao hydropower became vulnerable to environmentalist claims and to address concrete questions about the nature of those vulnerabilities. I have characterized uncertainty as a form of relation that constitutes late industrial environments. But claiming that late industrialism produces uncertain environments is a different kind of claim than saying that uncertainty comes to dominate the sociotechnical relations in question. How then does a highly unstable socioecological situation gain political traction? Through what technique does uncertainty cease to be merely a general condition perhaps unrecognized by the people involved, who may well be convinced they understand what is going on, to become the substance of a social experience and grounds for a practice? Activists' technical entrepreneurialism established sutures and connections that forced the industry into responsive but unpredictable configurations. Activists forced the uncertainties of life along Lao rivers into transnational networks of power and established a political ontology in which "environment" ceased to function as external nature or context to become apparent at potentially any joint in the delicate network of relations through which hydropower dams are built.

I examine the events of the activist campaign to demonstrate how the largely American activist group managed to be effective without the extensive popular organization of campaigns in India and Thailand and with no legal claims on the developers or the Lao government. In the process, I offer an examination of the practice of technical entrepreneurialism, as well as the related term "political ontology." Through what tactics, and therefore what political ontology, was their power constituted? That is, what reconfiguration of knowledge and nature made their endeavor both effective and surprising? And is it something to be taken seriously in the long term? What became interesting is the extent to which activists' whole raison d'être was predicated on a diagnosis of late industrial rivers, marking a creative ecological intimacy between activists, the hydropower industry, and the rivers in which both hoped for a certain kind of redemption.

As perhaps the most visible transnational antidam NGO operating globally at the time, part of IRN's story demonstrates a shift toward tactics of institution hacking. Institution hacking is a specific kind of technical entrepreneurialism, that is, an opportunistic (or entrepreneurial) mode of practice directed toward discrete technical relations by taking the opportunities and threats of those technical relations as its condition of possibility. If the term "hacking" is taken to mean the exploitation of material vulnerabilities of complex systems, then NGO activists are hackers of expert and decision-making institutions.

These tactics constitute a political ontology in that activists target material vulnerabilities of hydropower development of which the developers themselves are not fully aware. By construing these material relations in a particular way and enacting them as such, their tactics created a novel power relation latent in the materiality of the industry itself. IRN effectively created an experience in which industry actors felt vulnerable potentially at any point in the complex network of relations that enables the construction of large dams. Institution hacking is therefore a primary site for the production of uncertainty and, beyond merely a tactical shift, it entails a specific ontology of political possibility. By cultivating a mode of practice that views megaproject development as inherently risky, activists both demonstrated that risk and produced the institutional and discursive uncertainty that would later circulate through institutions and management practices. Activists did not simply mobilize public opinion or insist on a certain set of environmental goals or norms, and still less did they convince industry actors to agree with a different version of the facts. Yet they did not directly force the company to comply with legal obligations either.[3] Nor were they "working within the system" to try to make the best of it. Rather, they produced an experience of environment through the production of uncertainty. The result is a simple transformation. Under such conditions, environment ceases to be understood as context or nature "out there" and instead takes the form of permeating, unstable relations—an experience of vulnerability at potentially any joint in the network of relations that make hydropower possible.

If environmental activism serves to raise vital questions about the harms and costs of environmental events, we can begin to see in this chapter that it does so through a kind of emergent activity not usually recognized by studying the travails of controversy. This experimentation establishes sutures and relays that connect or bind ecological conditions of living to novel obligations. Obligations to rivers are far from necessary; on the contrary, they are difficult to establish and, as we will see, even more difficult to live up to. Reciprocally, the productive creativity of anthropogenic rivers is hardly the product of inspired minds with mysterious powers of creative imagination. Rather, to draw on Isabelle Stengers's (2010) phrasing, the suture of novel riparian obligations *forces thought* into responsive but unpredictable configurations. The suture, in a real sense, is more powerful than can be imagined at the outset.

Translocal Vulnerabilities

IRN's public campaign against the Theun-Hinboun Dam broke in March 1998, as the dam just began commercial operations. In conjunction with the dam's official opening ceremony, IRN released a press report detailing a range of environmental problems and villager dissatisfaction that stood out in strong contrast with explicit sustainability claims of the developers. The report was prominently covered in the Thai press and proved a major embarrassment for the Asian Development Bank (ADB), which had promoted the dam as a win for the environment. Subsequently, ADB sent a team of experts to investigate, which then published a mission report that revealed how little was known about the rivers and riparian livelihoods in question and effectively proved the activists' point for them. Through repeated follow up, IRN secured from ADB an open admission of the seriousness of the environmental problems and then pursued explicit promises for environmental reform through two years of consistent meetings and letter-writing campaigns targeting ADB. With the Theun-Hinboun Power Company (THPC) increasingly isolated, by the year 2000, the company promised to spend $10 million over ten years, beginning with establishing an environmental management division. Since the project's environmental problems originally were the legal responsibility of the Lao government, the success of this campaign resulted in not only the new monetary commitment but also the obligation on the part of the foreign-managed company to deal with the environmental problems on its own—a subjectivation of industry actors that, more than the extraction of a formal agreement, implies an experience of vulnerability. What could it mean for people in the industry to take these kinds of problems seriously? But there is more to this sequence of events than is immediately apparent, and, in order to think the contingency of these rivers, it is worth unpacking the tenuous, contingent power relations implicit in IRN's success.

IRN's office, located down the hill from the University of California, Berkeley, was housed among a complex of activist and NGO offices in a modest midcentury building behind an auto parts store. Although at the time I was pursuing my PhD studies at Berkeley, I had never heard of it until its name came up during interviews in Vientiane when I was trying to decide how to frame my research questions. IRN began in 1985 alongside two other notably formidable American environmental advocacy groups, Rainforest Action Network and the Earth Island Institute—the three of them formed at meetings held in the same San Francisco warehouse with considerable overlap among those attending (Mongillo and Booth 2001, 288). Indeed, these three NGOs would become well known for their savvy tactical disruption of corporate environmental degradation. These NGOs were entrepreneurs of the network form in the process of forging a novel array of environmental tactics.

Narrating environmental struggle often feels like describing a boxing match, but imagine instead a small team of activists working to connect collaborators through information and actionable agenda items for specific dam projects, in various states of development, all over the world, through the spatial form of the network.[4] Initially, IRN ran a clearinghouse newsletter, *World Rivers Review*, to catalog large dam projects and connect activist campaigns from around the world in response to the evident shift of large dam construction from Europe and North America to the rivers of developing countries. Founder Philip Williams was a UK-born and trained hydrological engineer who wrote technical criticisms of proposed dams to assist activist campaigns in the United States. As early as 1978, in the wake of the physical collapse of several large dams in the United States, Williams's technical criticisms underscored the unique kinds of uncertainties that accompany complex technologies and the sociotechnical calculation of acceptable risk. Because they involve novel issues of design coupled with site-specific risks such as earthquakes, he wrote, each individual "large dam involves essentially 'new' technologies. . . . Quantitative risk analysis approaches have been justifiably criticized because of the uncertainties involved in quantifying [risk], and because they restrict themselves to 'rational' failures when failure could occur in totally unforeseen ways and coincidences" (1978, 282). This kind of risk politics, recognizable as a reflexive subpolitics characteristic of the era (Beck 1992), over time would morph into an ontological politics through a tactics of disruption that sought to connect risk over literally disparate, physically separated spaces—from rivers in Laos to the hallways of the ADB. Whereas Williams's statement in this passage is concerned with the terrifying risks of low-probability but high-impact disasters—the idea of a dam collapsing upstream of a city during an earthquake, or the spectacular near-collapse of California's Oroville Dam in winter of 2017—the ontological politics characteristic of activists' later technical entrepreneurialism would center on "occupying" or hacking the institutional conditions of the industry, including its claims to truth.

With a staff of about ten volunteers and a paid editor, by 1988 IRN had made a name for itself by opportunistically holding a well-attended antidam conference alongside the annual meeting of the International Commission on Large Dams, an industry group, in San Francisco. It joined the protracted struggle against the massive Narmada project in India where it both infamously targeted the World Bank and drew criticism from some Indian and Adivasi (indigenous) activists for replicating the power relations of postcolonial transnational politics. Nonetheless, the minimal form of the network—a newsletter compiled with reports made by post; a mailing list; a circuit of contact and communication among allies, researchers, and contributors—belied a critical diagnosis:

> We saw the same obsolete big dam technology being exported to the Third World, disregarding the devastating ecological damage and huge economic costs that the U.S. had incurred in its dam-building boom in the 1950s and

> 1960s. Yet at the same time we saw successful examples of citizens groups' resistance in the early 1980s most notably in the cancellation of dams on the Franklin River in Tasmania, the Silent River in India, and the Nam Chaom Dam in Thailand. Around the world big dams were being promoted to gullible and corrupt politicians as a quick road to economic development by an international syndicate of self-serving interests: international consultants, construction firms, and development bureaucrats. . . . We believed that by cooperating with and coordinating with nongovernmental organizations fighting dams in international policy arenas we could counter the influence of these international influences. (Williams 1997 quoted in Khagram 2004, 184)

By connecting citizens' groups across the Third World, the nexus of politicians, experts, development banks, and construction firms could come into focus.

Ian Baird, one of the main activist researchers involved in the play of events in Laos, a Canadian, characterized his own biography in similar terms, indicating the links between localized events like the advocacy campaigns mentioned by Williams and the loosely connected, translocal networks of experience and tactics. Baird had come to work in Thailand through Earth Island Institute, which is well known for its aggressive tactics. His personal narrative similarly made note of a loose concatenation of localities. Before arriving in Laos, Baird offered,

> I was protesting that first dam, the Nam Chaom dam in Thailand that was the beginning of the Thai environmental movement in 1986. I worked with [Thai NGO] Foundation for Earth Recovery before it became TERRA, so I knew those people, and I worked with the NGO coordinating committee in Southern Thailand where I had been involved in activism against shrimp farms. . . . All of the activism in Laos was intended to target the Asian Development Bank, not the government—to challenge the company, not the government. In retrospect maybe that was a mistake but that's the way we thought about it in those days. We thought the government was being duped by all these big companies, they were being drawn into these projects where they didn't know what was going on. We thought we could convince them.

Before he came to play a key role in liaising with IRN in Laos, Baird held a literal connection to the IRN's milieu, an analogous itinerary through disparate local campaigns and a tactical commitment to target multilateral banks and multinational corporations as a matter of articulating care, over developing countries presumed ill-equipped to deal with globalization.

IRN came to work in Laos through a cluster of primarily expatriate NGO workers and activist researchers, including Baird, who worked at inconspicuous development and community volunteer organizations. Sombath Somphone, the prominent Lao human rights activist who was kidnapped and disappeared for his activism in 2012, was one of the few Lao people involved. The organizations where these people were based included Japan Volunteer Services, the Canadian volunteer organization Cuso, World Education, and Community Aid Abroad—in other words, neither activist groups nor environmental conservation groups. From the vantage of activists based in Laos, IRN worked through them, following their advice and relying on them. The production of knowledge about Lao rivers, Lao communities, and hydropower projects was a key element. By the time of my research, IRN's California-based staff characterized their strategy as documenting the concerns of local people in order to raise those issues in higher profile settings—specifically the institutional channels of multilateral finance and diplomacy whereby they targeted executive board political representation as citizens of development bank donor countries. This approach of taking details of what is happening in villages and along rivers and then inserting those details into high-powered institutional settings should be understood as a specifically spatial strategy.

The Laos-based researchers drew powerfully from Thai environmental thought, which over several decades had developed a deep philosophy of community organizing quite different from the comparatively militant tactics of IRN.

As Baird described it,

> In Thailand I was employed by the Earth Island Institute, founded by David Brower. I started working with them with a typical North American environmental view point that I had gotten when I was pretty young. . . . Earth Island was using US legal instruments to force other countries to do certain things, they were interested in using US trade power—they were somewhat grassroots on the US side but they were coming out of San Francisco with that noncompromising view that the real pure environmentalists are the ones who don't compromise. . . .

> I became very interested in how the Thai NGOs were working in a much more social way, with much more grassroots participation that came out of the communist party in Thailand. It wasn't environmental imperialism at all. I came in conflict with Earth Island Institute and I ended up leaving with some major differences of opinion. I had fully converted to this idea from the Yadfon Association in southern Thailand—the whole idea was to let people try to figure things out for themselves and you try to support them in whatever way you see fit. . . . So I was modelling what I was doing in terms of the principles, not the actual details but the principles, of what had been done in Thailand. You could call it ecological Marxist if you wanted a term—I don't think I used that term at the time but I think that would describe the Thai philosophy. Anti-capitalist, pro-villager—that was the position I was coming from then.

In fact, the Yadfon Association is well known for protecting coastal mangrove ecologies through an integrationist approach with shrimp farmers, an industry exposed to tremendous market pressure often with devastating ecological and labor consequences (Stonich and Vandergeest 2001). Its founder, Pisit Charnsnoh, was awarded the prestigious Goldman Environmental Prize in 2002.

This strategic orientation has largely been characterized as transnational activism—as indeed I have described—but it is also worth characterizing its spatiality as *translocal* to the extent that environmental campaigns were linked through analogous specifics in different places. Comparable vulnerabilities and tactical experiences enable people in diverse locales to articulate a "pragmatics of struggle" (Pignarre and Stengers 2011, 17–19). Pragmatically, it is counterproductive to view institutions as totalizing and closed, for that reifies a distinction between inside and outside of an abstract system and gives the institution more apparent integrity than it deserves. Conversely, inserting these experiences of harm into the networks of finance, planning, and project approval can serve to force thought within spaces of power otherwise insulated from knowledge about those experiences. As a spatial strategy, it draws into relief the delicate separation of ecological risks from the procedures and institutional pathways of decision making, while emphasizing the porosity of those procedures and pathways. The institutions of expert development planning and finance serve to externalize environmental impacts; activists' attention to material modes of networked connection served to undermine that constitutive exclusion and multiply threats for hydropower development.

This style of environmental activism, which emerged in conjunction with Southeast Asian environmental praxis, is effective for four distinct reasons. It can forge lateral relays between localities where campaign experience and tactics can be collated and dispersed. It articulates demands in the personal register of lived environmental damage and precarious vitality of communities, drawing on discursive resources that bridge the technical and the affective. It taps into core themes of capitalist and state disenfranchisement without, however, taking the form of a critique of an abstract system ("capitalism"; "the state"). Lastly, it bears on the comparatively plausible goals of shutting down or forestalling given projects, winning compensation or extracting entitlement compromises. Because of its connection to lived experience and comparatively specific, local targets, environmentalism in Southeast Asia has become an important domain of political organization. For example, the populist opposition movement in Thailand, whose rise to power in the early 2000s later prompted a series of military coups, had at least one important point of origin in rural activism against the Pak Mun Dam (Missingham 2003). A crucial element, therefore, is the lateral transfer of knowledge and experience from one project to the next, which depends on diagnosing the riparian environment.

Riparian Diagnostics

While strategically organized, the campaign against the company was circumstantial and opportunistic, if ultimately provocative. Bruce Shoemaker was hired by International Rivers Network in 1997 to research and write a report on hydropower and Lao rivers—a report that would later be published as "Power Struggle: The Impacts of Hydro-Development in Laos" (IRN 1999). Too many projects were being proposed to keep track of, and it was very difficult for activists to get details about their potential impacts. "The idea," he said, "was to show that all of these projects were being developed in terms of the lack of acknowledgment or consideration of local livelihood issues or issues of what happens to the river, the poor practices for resettlement." In particular, "there was very little information or acknowledgment of fish migration issues in the international NGO community, much less the wider development community." He had come to Laos with the Quakers and, with a commitment to social justice rather than any specific environmental issues, had worked on a range of peace and reconciliation projects such as unexploded ordnance removal. However, through the years he had come to collaborate closely with Baird to produce research on riparian livelihoods and potential impacts on fisheries. Having lived and worked in Laos for seven years—at the time, far longer than most expatriate development workers—and being fluent in the language, for the IRN report, he worked with a Lao colleague (who remained anonymous) to gather information from villagers to be affected by a number of planned dam projects.

Working with Lao colleagues was delicate, for challenging a state project was clearly a dangerous proposition and the question of who can speak was central to the legitimacy of environmental claims. Lao colleagues had worked on that IRN report and others but deliberately remained uncredited in order to protect them. Reports were anonymously translated into Lao by sympathetic development NGO staffers. Other Lao working on community development projects played pivotal roles while not necessarily fully being brought into the activist circles and potential controversies that might surround their involvement. Somphone, the human rights activist who was recently kidnapped, was an exception. He worked closely with transnational activists and chaired one critical meeting between IRN and executive directors of ADB in Honolulu. Shoemaker told me, "We always wanted to have Lao people and Lao voices involved but not wanting to get people in trouble. Sombath and I talked many times about that issue. And his joke was always that 'foreigners would get kicked out but Lao people would get kicked in.'" Even those villagers whose complaints were documented in IRN reports were put in jeopardy.

In preparing the report, the Theun-Hinboun Dam, which was completing the testing phase just after construction, was the only operational project—the first of many to come. Its official opening ceremony was planned for just a few weeks away, and they decided to release a short media report detailing his findings from villages within the impacted riparian zone. As Shoemaker recalled, "It was a fairly spur of the moment decision. . . . The opportunity was obvious with the ADB president coming to this big opening ceremony in Laos. There was an opportunity to be much more high profile and get this information out right away at the time of the opening. The [dam was] being portrayed as a straight, new opportunity—an environmentally-friendly dam. I think it was two weeks [of preparation] probably. . . . Now, I didn't know it was going to be on the front page of the *Bangkok Post* when it came out. It was higher profile than I imagined it becoming." This opportunism was corroborated by archival emails I was able to view.

Environmental experience in Thailand once again played a pivotal role. "We smelled a rat. The World Bank had [similarly] built supposedly a run-of-the-river dam in Thailand with a small reservoir, but yet people were having all these impacts up and down the Mun River. It seems a little bit strange that the ADB could do the same in Laos and everything was just fine." A run-of-river dam is one for which the water level supposedly does not exceed the original river channel. The environmental impact assessment for the Theun-Hinboun Dam was predicated on the idea that the dam would not flood more than the existing, natural channel of the river and therefore the impacts would be minimal, with zero resettlement. In fact, the environmental consulting team had not even bothered to collect baseline data about socioeconomic conditions; its environmental assessment read as a checklist with item after item dismissed out of hand with no empirical backing, and its recommendations made blithe claims that the dam would be a major benefit to villagers' lives.[5]

In the context of building a long-term research base addressing environmental impacts, the activists were situated to grasp an opportunity made all the more apparent by the discursive parameters set up by project developers. Thinking they were insulating themselves against criticism, the developers excluded even minimal field research from their reports, leading to a lack of data that would come back to haunt the company. Through their seamless positive spin on the project, they made it all the more apparent where their vulnerabilities could be found. Moreover, activists' commitments in Thailand had given them a wealth of tactical knowledge, whereas the developers assumed that the lack of organized civil society in Laos would guarantee the privileged authority of their expertise. Such an entrepreneurial tactic on the part of activists does not result simply from releasing a report critical of an industry, no matter how well researched. Its circumstantial opportunism rather comes from the technical (if partial) awareness of an industry's functioning and its limitations. They had diagnosed the developers' naïveté, the industry's organizational ignorance, and its practitioners' overall lack of reflexivity.

Within that network diagnosis, however, was a more fundamental problematization of extensive ecological problems and a claim that they should be problems for the company. When I questioned Shoemaker about the extensive criticism of his research from people who felt his claims were unfair or biased, he responded by pointing directly to the problematizing dimensions of his advocacy work, in stark contrast to essentializing modes of environmental discourse that have elsewhere been noted: "[My research] was simplistic in that it was a very short study based on interviews in a few villages. On one hand, I think I was completely truthful. I did not try to only bring up the negative voices. I did not hear any positive voices other than from the Theun-Hinboun office. It was overwhelming, everywhere I went, as I described in the report, people were talking about the impacts on the fisheries. But I didn't mean it as a comprehensive fisheries study, but more as a callout that it looks like there's a big problem here and we need to have some comprehensive fisheries studies. Because from this short, unscientific sample here, it looks really bad."

Activists worked to catapult the "event" of the dam into those specific networks that deserved to answer to its problems. Within the right assemblage of practices, to draw on the language of Isabelle Stengers, the event of hydropower *forces people to think* differently about riparian ecologies. Only through these diagnostic and assembling practices did the river become anthropogenic within these spaces and institutional relations.

Part of my claim is that these truth claims do not so much attempt to produce another version of authoritative or essentializing knowledge, but rather take up an incessant problematizing or diagnostic role of articulating the voices of people whose experience gives them grounds to speak. To take only one instance, in the words of one of his interviewees: "Now it is very difficult for fishing. We can only get about half as many fish as before the dam closed. We don't know where all the fish went. We have to buy expensive new nets to try to fish deeper in the river now. It is very difficult. Also, the vegetable gardens along the riverbank have all been flooded. They were each worth more than 100,000 kip [about $40 in 1998] but now are gone and much erosion has occurred. There are many problems. We have not been told about receiving any help or compensation for these problems" (group of fishermen, quoted in Shoemaker 1998, 8). Erosion, loss of the fisheries, loss of fertile riverbank soil, and collapse of the annual patterns of flooding were all essential questions tracking the emergent uncertainty of the anthropogenic river. Articulated as complaint and demand for better treatment, these statements, however thinly captured in the activists' report, bear directly on the ontological status of the river—a question of water that is too muddy to fish, of unpredictable and excessive flooding, siltation of key ecological areas—in short, not simply a matter of the destruction of a river but the creation of a novel ecology whose parameters are not simply different but, for the foreseeable future, continually in flux, with vital implications.

Riparian diagnostics hinge on an essential question—what is this river, and what has it become?—and many of the most significant people in the activist campaign were activist-researchers who, in various ways, were directly interested in the socioecological viability of wild capture fisheries. Riparian fishing organizes a gestalt of regional human ecology and, because it is poorly suited to capital intensification or development intervention, its routine near-absence in debates about development in Laos signifies a decisive blind spot in the program of modernization. If fisheries are the target of development, it is usually with respect to aquaculture and fish farming (but these are difficult and rarely taken up). Rice farming by contrast is frequently emphasized by agricultural development projects because it seems ripe for capital intensification, marketization, and the application of expertise. If ethnic Lao livelihoods have long been organized spatially along rivers, they also have been organized culturally and temporally with respect to seasonal flooding, fish migration and spawning events, seasonal rice agriculture, and the cultural technologies that ground a *longue durée* social ecology. Development practitioners routinely identify Lao villagers as farmers; these activist-researchers insist instead that rice cropping is only one part of their livelihood strategy. Advocates such as Shoemaker, Baird, or Maiveng, the Lao filmmaker, emphasize wild fisheries because fish bear directly on the culturally specific, frequently pleasurable, and comparatively politically independent lives of people struggling in the face of powerful, modernizing forces.

Moreover, many of these same activist-researchers play an essential role of engaging in peer-reviewed research for underresearched empirical topics, providing expert consulting services in cases where fisheries expertise might have an important impact, and providing essential input for activist campaigns for dams across the region. Terry Warren was one prominent fisheries expert who had taken an important contract with the THPC as it tried to respond to its critics. When the company refused to accept his conclusions, he circulated his research independently and spoke out about the dam's impacts. Another, Steven, whose role I explore further in chapter 4, rose through extended activist work around the Pak Mun Dam in Thailand before taking on a commercial hydropower consultancy that ended badly, with him going on to collaborate with IRN. He later enrolled in a doctoral program in the United Kingdom. Baird was trained as a social scientist but spent much of his career engaged in community development for wild fisheries, with significant published research bridging ichthyology and social sciences, before taking a tenured job at a major US public university (the recently discovered fish species *Schistura bairdi* is named after him). This activist-scholar and sometimes-consultant bridging role is tied to an affirmation of local riparian lives that expresses itself in careful attention to the health of fisheries.

A particularly striking instance of this affirmation of riparian life occurred during my research when the hydropower company funded a major fisheries study that was methodologically grounded in interviewing fishermen about their catch, and in some cases hiring fishermen to collect systematic data, as a way of assessing fish populations. Steven, hired to complete some different work, homed in on an obvious methodological limitation: the survey was focused on trying to figure out the would-be objective state of the fishery, rather than the socioecological state of the riparian ecology more broadly, by focusing on large fish caught usually by male commercial fishermen. By contrast, Steven emphasized "living aquatic resources" rather than large commercial species by pointing out that the wet season ecology produces all sorts of living things, many of which are spawn, minnows, or small crabs, caught by children or women fishing not in the main channel of the river but along small streams and seasonally flooded areas. For Steven there was a diagnostic connection between the gendered interest in large fish, the focus on fish numbers rather than ecology, and the exclusion of the diversity of aquatic resources relevant for people's use of the riparian environment.

When caught, this much broader array of smaller fish and spawn will be layered with salt and fermented into a thick, pungent, protein-rich fish paste, *padaek*, which will provide an essential nutrient source for the coming year. The aquatic biomass (including larger fish) forms the bulk of villagers' dietary protein. Yet hydropower serves to disrupt precisely the temporality of annual flooding so essential to migration and spawning events. Seasonal streams turn into muddy backwaters. Below the powerhouse, the temporality of flood and ebb become tethered to power demand in Thailand, while the persistent, unending erosion completely transforms the morphology of the river. Of all the many problems caused by this dam, the radical impacts on the fisheries through the transformation of the riparian ecology will remain the most significant impact. But this diagnosis is inseparable from a form of activism that insists on a certain domain of problems and builds the real relations through which those problems gain publicity and traction. If there is no independent access to an ontology of nature, we find the essential question "what is this river?" taking form through the demands and diagnostics of a translocal activism capable of raising difficult questions for an industry structurally unable to answer them. This form of activism instantiates what I have called technical entrepreneurialism.

Technical Entrepreneurialism

Shoemaker's report was picked up by the major Bangkok newspapers and was covered alongside the opening ceremony of the hydropower dam. Juxtaposed against ADB statements that "there is little for the environmental lobby to criticize" (ADB 1997, 8), and major fanfare about innovative financing structures at the very height of the pre-1997 Asian investment boom, the activists' report was a media event in the making.

Six years later, memories of these events were still acute, as expressed to me during field interviews. Many in the development circles in Vientiane singled out the public nature of the criticism and the ensuing controversy, remarking that activists had taken the wrong approach; that their criticism was not wrong but they had been unnecessarily antagonistic; or even that they had betrayed the trust of Lao partners whose primary concerns were development. Government officers I spoke with said they had been taken aback by the public criticism and that the issues should have been raised privately. A Bangkok-based hydropower financier reported that he had to reassure his Lao colleagues that this "aggressive approach" was "the Western way" of dealing with issues, and told me that, "like children," government staff wondered why they were being criticized when they felt they were doing everything right. The head of a major NGO, a British woman who had once given technical input on the dam as a consultant, told me "[Shoemaker's] report was too strong."

Government officers focused on the Thai media coverage. When I asked one government technocrat what his reaction was at the time, he leaned back and laughed, taking his time to formulate a diplomatic response. He said, "It's about 50-50. I'm grateful for that report. The impacts were serious and it was good that someone brought them up. But at the time everyone was angry." The environmental manager of another hydropower project, an anthropologist, told me that the report had proven important in raising environmental concerns within the industry, while another group of consultants subsequently developed a back-channel relationship with IRN in order to leak controversial information. The report's rapid uptake proved a welcome surprise for Shoemaker and staffers in IRN's California office, and Shoemaker himself was well aware of the loss of trust that came along with his criticisms: "for some of my colleagues in the Lao government, people I knew in the Ministry of Foreign Affairs, it seemed like sort of a betrayal. They were kind of stunned that I had authored something that was so critical of the situation in Laos."

In response to the report, ADB immediately fielded an aide mission, ostensibly to investigate IRN's factual claims. But the tone of an antagonistic battle over facts was already in place. Quickly the hydropower company and ADB took an entrenched position against the activists' claims of damage, with the effect of stoking the controversy. Beyond the initial field research, publication, and media activism, the outcome of the campaign was heavily dependent on systematic, labor intensive follow-up by transnational activists able to make direct demands on institutions outside of Laos. By late 2001, after a barrage of reports and counterreports, volleys of letters sent to dozens of potentially interested parties, multiple fieldtrips and aide missions, and meetings in Manila and Honolulu between activists and senior executives of the ADB, ADB eventually accepted many of activists' concerns and forced the company to accept expanded liabilities.

Francesca Polletta, a social historian of the US civil rights movement, has described the mundane labor of participatory democracy with the strikingly unromantic phrase "freedom is an endless meeting" (2004). While the IRN campaign had the superficial appearance of a media event, it is no underestimation to say that the demand for "free rivers" was an endless email exchange. Much of that mundane labor of institution hacking involved both detailed knowledge of institutional relationships and technical understanding of the many vulnerabilities of hydropower planning practice.

I analyze this labor to demonstrate institution hacking as a kind of technical entrepreneurialism, that is, a set of technically sophisticated practices oriented toward unexpected manipulation of specific material-semiotic relations and for which opportunity and threat form an explicit condition of possibility.

The subtext of activist correspondence demonstrates the kinds of relationships in play. These letters can be read crosswise as ethnographic artifacts, as I do for a consulting report in chapter 4. Here is a brief selection of their exchanges.

ADB, responding to IRN's ongoing documentation about environmental damage: "We [ADB] therefore do not agree with your allegation that ADB initially refused to acknowledge Project-related social and environmental impacts, and only by late 1998 acknowledged for the first time such impacts. In fact, it was always known that Theun-Hinboun, as any other project of this nature, will have environmental and social impacts, which will have to be mitigated."

IRN, responding to ADB's January 2001 response to IRN's letter: "The response reiterates the same distorted claims made by S. Bauer in the May 1998 ADB Mission Report. . . . 'Trouble on the Theun-Hinboun' did not state that there was an 'emergency situation.' This is the phrasing the ADB itself used when trying to discredit [the] initial findings as being exaggerated. . . . Three years later . . . it still deserves the ADB's immediate attention and I am extremely disappointed at the unjustifiable delays."

Request from a Swedish Journalist: "Dear [IRN], I am very interested in the following reports [by the ADB]. . . . Do you think you could help me find these reports and send them to me . . . by ordinary post? People in ADB are very unwilling to give me copies. [Signed]."

A Norway-based activist, writing to IRN, referring to the Norwegian development agency: "In my view, IRN is at a real disadvantage sending this letter because you will not be able to do the lobbying and footwork required to force a response from NORAD [Norwegian Agency for Development Cooperation], and the government is under no obligation to be accountable to a foreign [environmental] group."[6]

For the campaign, IRN worked closely with a small number of other transnational groups, including FIVAS (Foreningen for internasjonale vannstudier, Norway), Probe International (Canada), Mekong Watch (Japan), and Terra (Thailand). All of these groups drew directly on their ability to make claims on their respective national governments as an essential element of their own environmental citizenship. FIVAS, the Norwegian NGO, was very active campaigning with Scandinavian shareholders and development lenders. Correspondence included email addresses of delegates who had conducted site visits, details from phone calls with Statkraft representatives and the president of Nordic Hydropower (both shareholders), and off-the-record reports from internal sources. For IRN, it meant direct lobbying of the ADB executive directors, with leverage through US political representation through a senior advisor to the US Treasury.

The activist network systematically worked to undermine the institutional process by inserting themselves into key nodes or relationships, especially those pertaining to authoritative knowledge. While most of their activism concerned social and environmental knowledge, they also published critiques of the economic rationale of the project, sought economic analyses of specific dimensions of the hydropower plans (effects of upstream dams and minimum downstream release on revenue), used environment as a measure of ADB's accountability, and even issued sector-wide critiques of energy demand forecasts in Thailand used to justify purchasing agreements. Later, IRN would go on to publish research reports on the structure of hydropower financing, including the role of export-import banks and other esoteric financial arrangements. Their work was therefore much broader than just environmental criticism; it rather exploited a wide range of vulnerabilities to produce uncertainty within the otherwise stable power relations that enable hydropower development.

To describe one instance of how distributed letter writing functioned, the hydropower company, THPC, had hired a fisheries expert to produce a comprehensive report detailing the state of the fisheries after beginning operations of the dam. This fisheries report had been substantially delayed, and activists suspected its conclusions would provide important data for their campaign. Through one of its contacts in Laos, IRN came to understand that the fisheries consultant had a major disagreement with THPC, yet the consultant would not say what the issue was precisely. IRN argued to a senior advisor to the US Treasury, who was in turn in contact with an executive director to the ADB, that ADB was going to water down the report. This executive director could then put pressure on the ADB through their civil society office (which at least some people felt was created in response to this particular activist campaign). This communication was meant to improve the negotiating position of the consultant so THPC would stop stalling in releasing their report (naturally without ever explicitly mentioning the dispute itself).[7] This kind of distributive communique shows how activists developed intimate knowledge of existing institutional channels to put them to work for their own purposes.

A more common tactic was to submit a letter of demands or requests for information, then to rely on a large number of allied groups to write follow-up letters to ensure a response. In one notable case, the initial letter of demands went to the president of the ADB, copying all the ADB executive directors. The IRN campaigner then reached out to a sizable number of other NGOs to write follow-up emails stating that the response was inadequate. In this case, the list of NGO allies included nineteen organizations.

The Norwegian NGO FIVAS created a detailed map of project institutional connections through which the dam was financed and built (figure 2.1). While environment was essential to this configuration, the technical knowledge used to support their aims far exceeded environmental knowledge narrowly construed. Two things are remarkable about this mapping labor. First, *information gleaning*: the level of detail is not simply a replication of information easily available in existing form, but rather required extensive gleaning of facts from dispersed project documentation and unofficial resources. The diagram is not a model of the hydropower industry as a generality, but a model of an instance of its conceptual specificity. Second, *uncertainty displacement*: the outcome of the systems risk analysis, in the lower right of the diagram, is that hydropower development involves a major risk "subsidy" when local communities and the people of Laos absorb all the ecological risk of the project. What hydropower offers to the public is little benefit but much risk—in their word, "uncertainty" (far bottom, bottom right).

Mapping of risk relations bears on socioecological aspects from a distinct vantage point. An email exchange between FIVAS and IRN showed how rainfall risk played a role in project vulnerabilities: "[Our government source] confirmed (*off the record*) that Theun-Hinboun is insured with Export credits both from NDB [Norway Development Bank] and ECA's [Export

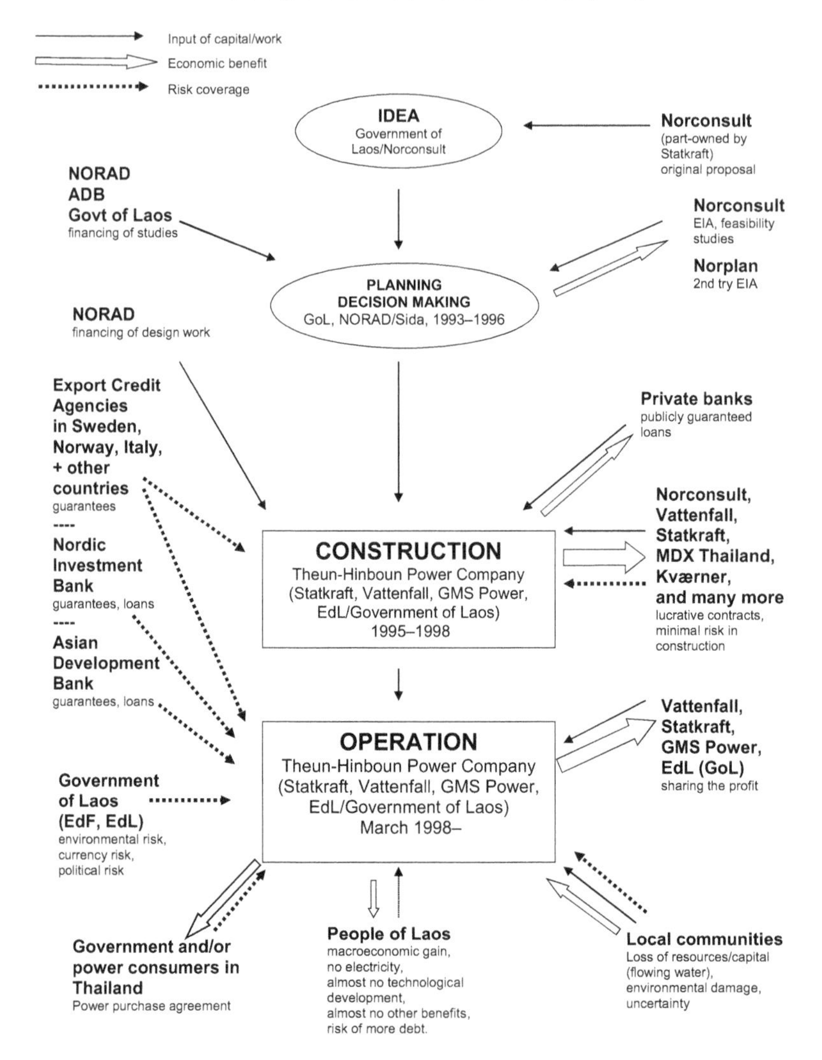

Figure 2.1. Conceptual map of THPC's global risk assemblage, discovered in an archive of activist materials. Diagram by Øyvind Eggen/FIVAS, used with permission.

Credit Agencies] in Norway and Sweden. She wouldn't give any figures, but she said that the export guarantees had taken into consideration both possible changes in currency rates, possible reduction of rainfall and other technical difficulties." Related correspondence addressed tricks for gleaning information; for instance, in one case, a researcher notes that hints of otherwise confidential information can sometimes be found in promotional brochures; in another, they shared insight for determining the legal limits of liability for public institutions in different countries. The degree of technical precision is notable.[8]

There is an implicit parallel between the rationality of networked connections, in which every moment in a circuit of tenuous links requires maintenance and may become subject to failure, and the contemporaneous logic of risk in which apparently solid systems can be quickly shown to be fraught with vulnerabilities. As Bruno Latour argues in a commentary on Ulrich Beck, risk "does not mean that we have run into more dangers than before, but that we are increasingly entangled, whereas the modernist dream was to disentangle us. . . . A perfect translation of 'risk' is the word 'network' in the ANT [Actor Network Theory] sense, referring to whatever deviates from the straight path of reason and of control to trace a labyrinth" (Latour 2003, 36). Activist labor mapped this labyrinth of institutional vulnerabilities.

Activists shared in the spirit of the times with theorists such as Beck and Latour, and indeed their work transitioned from the reflexive politics of the former to the ontological politics of the latter (see also Woolgar 2005). Their work demonstrates the technical entrepreneurialism through which diffuse, local, and externalized environmental risks are made to function as so many vulnerabilities within the material-semiotic circuits of hydropower planning and finance. Technical entrepreneurialism is an inherently relational modality of practice that works by opportunistically assessing the potentiality of real relations in order to explore and exploit. Reading this institutional knowledge practice as a kind of materialism shows that activists deploy a political ontology whereby specific, real relations, which are not divided into categories of natural, technical, or social, can be exploited in ways unexpected by those whom the institutions are meant to protect and sustain. In doing so, activists tactically exploit the manifest potential of all sorts of effective relations (documents, email, industry vulnerabilities, specific expertise). As a practice, technical entrepreneurialism implies attention to technical detail or the inner workings of technical systems and institutions; yet "technical" in this sense is not the opposite of "political," but rather alludes to relations' specificity, their status as delicate achievements, and the obligations they place on those who build, maintain, or deconstruct them. An ideological critique is insufficient; the activists must create for themselves an intimate tactical diagram of the system's weaknesses and turn the power of the arrangement against itself. The kind of knowledge deployed depends on the detailed understanding of institutional and technical systems' practical dynamics, for it is these practical dynamics that are opportunistically disrupted. This approach in turn relies on a very different understanding of institutions than the black-boxed monoliths that tend to characterize ideological critiques. Attention to the vulnerabilities of technical systems gives an ontology of systemic vulnerability.

Every politics concerned with its own effectiveness requires an ontology as a theory of its own nonsubjective powers. "Political ontology" is a term used by Helen Verran (2014), Anders Berman (2017), and Mario Blaser (2013) to describe an ontology with political stakes, much in the sense that Martin Holbraad, Morton Pedersen, and Eduardo Viveiros de Castro (2014) articulate a politics of ontology in terms of the "permanent decolonization of thought." That is, their use of the term designates that ontological pluralism is the political program of their scholarship. Although this work is allied with that project, my use of the term is different, for it designates how an actor or group, not necessarily human, ensnares others within a political strategy based on a specific set of ontic commitments, obligations, and judgments regarding the other. Thus political ontology is not only an interested orientation toward what could or should be, or a conflict about or between different realities (Blaser 2013), but is also an evaluation, judgment, or decision about what exists as a theory of powerful action. It is a calculation or gambit about what kind of world is possible and what strategies of power may be effective. Furthermore, this evaluation implies a commitment, that is, an interested, risky investment in a power strategy inevitably with its own ontic consequences and vulnerabilities that cannot be fully known in advance. In my view, it is this tension between evaluation (which is a normative activity) and commitment (which is subject forming and constitutively transformative) that forms a critical element of political ontology, more so than conflict or power per se.[9] Politics is inherently evaluative, affirming and normative;

it is ontological when it affirms or denies matters of existence (including matters of life and living).

Activists forced the company to contend with environmental concerns not by constructing an external, reified nature, but by making the industry subject to pervasive vulnerabilities at potentially any joint in the complex relations through which large dams are financed and built. In doing so they made an industry heed some of the problems it had caused, not by convincing it of the validity of their concerns through superior reason, not through a scientific epistemology of fact-making or by enforcing the law, but by making business subject to a labyrinth of unknowns.

A Labyrinth of Unknowns

Deconstructing authoritative knowledge was the upshot of the activist work, precipitating a crisis of expertise and giving form to the mess of uncertainties that came out of the hydropower project. Page-for-page, technical criticism made up the bulk of the correspondence. Again and again, disputes over questions of fact resulted in a wealth of embarrassment for the company and the ADB, which were put in the position of demonstrating that they had little factual understanding of the riparian situation. Passages quoted in the text box indicate the barely restrained rancor when it came to the legitimacy of activists or the industry to make claims of fact. But whereas activists were in the position of asking difficult questions and showing how the ADB's own rules had not been followed, industry actors frequently made obviously unsubstantiated claims, took a tone of dismissiveness, and relied on judgments that demonstrated their own invested bias. In acting, they demonstrated who they were. Because industry actors adopted the discursive position of relying on authoritative claims of fact, activists' problematizing stance served to unravel the assumptions of their discourse. Moreover, as we have seen, activist tactics also forced a shift from the discursive to the material, from claims made about Lao rivers to the impoverished instruments that made possible that objectification. The momentum of this ontological politics comes from the effect of turning the power of the mechanism back onto itself.

Claims by the industry were latched onto by activists to reveal the paucity of empirical evidence and the tendency to bully villagers. The initial ADB aide mission report was the most striking (ADB 1998, n.p.). In one passage, the aide wrote, "In the opinion of the Mission the village is located on the banks of a new beautiful large lake." Another passage argued that villagers' could no longer catch fish simply because the water was muddy and "the fish can no longer be seen." Numerous passages shifted the blame to villagers—citing illegal fishing, primitive equipment, or population pressure for the sudden inability of people to procure a livelihood. Statements like these, which were astonishing by standards of careful sustainability-speak in Vientiane only a few years later, only showed the stark naiveté of complacent (and frequently personal) connections between development finance and project developers. They also sparked a crisis of expertise because they rest on claims about the company's factual responsibility and yet reveal there had never been adequate research to understand the potential socioenvironmental impacts of the project.

Jay Daley, a Canadian certified accountant and former long-time ADB staffer, described to me how he became Theun-Hinboun's manager, after the ADB had financed it. "I had been working in ADB itself for I guess for 14 years. . . . I thought I was going to retire, but 10 days before I left, Bob Kay, who had worked with me at ADB, came to me and said 'would you be interested in running this power company in Laos?'" Robert Kay, also once a project officer at ADB, was involved in issuing the original environmental documents before going to work for the prominent Thai shareholder, GMS Power, where he later rose to executive vice president.[10] I interviewed Daley long after his retirement over the phone from his home in British Columbia. He had been difficult to locate; I eventually made contact through a social network of retired ADB employees.

Daley suggested the company and ADB maintained a strong relationship throughout the struggle with activists, but it seems clear that ADB eventually soured on defending him and the dam. In correspondence with IRN, ADB described negotiations with Daley as having "limited results" and as "mostly negative." Fisheries were "intensely discussed" before a moderate resolution was reached. Daley's version: "I was told by the board, spend whatever you need to do, just get this problem to go away. That was effectively what it was. It was more specific advice than that. That was the net-net of the whole thing, make this problem go away." ADB's support was substantively challenged by the activists' campaign.

One event stuck out in people's memories, in several cases more than a decade after the fact, including Daley's. The cover of the original activist report included a photo of a villager standing on the veranda of a house, touching a fishing net suspended from one of the house posts (Shoemaker 1998). The report unthinkingly included the name of the man's village. Wearing a floral shirt and a black jacket, the man was quoted as saying "The water levels this year since the dam was closed are the lowest we have ever seen for this time of year. The fish have all fled to the Mekong." A month after the release of this first activist report, company managers and a government operative (and ostensible communist party member) identified the villager and went to track him down.

Daley, the then-general manager, described the event like this:

> There was a big article about us having destroyed the fishing industry in this particular [river]—totally blew us away. I got my guy. I had a Lao administrative guy who . . . was a very sharp guy. He was an electrical engineer and educated in Moscow. A Lao, but a very, very capitalistic Lao [the ostensible party member]. We went down to find this place because we knew roughly where the [fisherman] had been talking because of the description in his article. We knew roughly where it was. We came across these guys. They were sitting under a house drinking beer. And we sort of rolled up and said, "Hi, how are you doing, blah blah," and they offered us some beer, which is a good sign. We started talking to this guy and he said about the fishing—no fishing. And he said, "No, of course there's no fish coming this time of the year. It has always been like that." [IRN] represented us as if we destroyed the fishing industry, but that particular time of the year, there was no fish and there never had been, so we didn't do anything about it.

Without too much exaggeration, it is safe to say that this caliber of evidence is what the company based many of its claims on.

Following this encounter, ADB's aide mission published another photo of the villager, wearing the same shirt, standing on the same veranda, sheepishly holding the activists' photo of himself, along with an apparent retraction of his criticism of the dam. Oblivious to the image campaign being waged against them, industry representatives effectively published direct evidence of their simpleminded and abusive attempts to sideline any concern for villagers. The fact that Daley still understood the event as vindicating the empirical issue of fisheries only underscores how ill-equipped the industry insider was for thinking about ecological issues.

The fact that the project developers were singularly unprepared for public scrutiny created a kind of hallway of mirrors. The company demonstrated its own vulnerabilities by looking like a predictable representative of a notoriously damaging, engineering-focused industry.

Initially, the logic of the company's environmental reasoning was strictly causal and legalistic: what effects can be shown to have been caused by the impact of the dam? What contractual agreements has the company entered, and what are the legal requirements for it to revise its approach? "A lot of issues boil down to law and the interpretation of the law," wrote a consultant to the project at the time. "The gov't signed a license agreement with the THPC before a lot of the critical environmental and economic issues were fully addressed. . . . They work strictly within the limits of their legal agreements." (For her part, she preferred to remain on the sidelines, in spite of her social impact concerns, cautioning that "one never has access to all of the information all of the time.") But the legal basis for the company's contractual security ultimately could not take into consideration the totally unexpected kinds of relationships that were being undermined. Daley did not care to elaborate very much, but he did say, "Initially, we were looking at the legal route, but then we decided the legal route wasn't the way to go," in order to protect the company's reputation.

This sharp affective dimension of the activist campaign, over the course of those three years, progressively isolated the company regardless of its legal protections. In breaking the cozy relation between the company and the ADB, the activist campaign produced a serious shift in the operational assumptions the company had been able to take for granted. The terrain had shifted from battles relying on authoritative knowledge claims and insulated, bureaucratic modes of expertise, to a terrain of media and diplomatic struggle emphasizing images, reputation, transnational networks, and the rancorous affect that broke apart the security of building dams in politically remote countries and insulated institutional networks. These experiences are essential to understanding the performative, media-savvy, and frequently dissimulating managerialism that has come to characterize the corporate environmentalism of Lao hydropower, which I take up in chapter 3. Just think of the incredible time requirements of incessantly responding to each email query and every insinuating riposte. Generally speaking, these experiences map onto three domains—image, affect, and network—which became part of the crucial terrain of postnatural, anthropogenic rivers in Lao hydropower development.

These effects were not limited to the hydropower project in question. By the time I arrived in Laos to begin my ethnographic work, the prominent Nam Theun 2 Dam, the World Bank's model project discussed in chapter 1, was in advanced planning stages. Its planning process had become a hyperbolic expression of planning taken to its logical extreme. By simply demanding that environmental planning adequately take into account socioecological complexity, the very logic of rational simplification turned against itself to produce a mimetic profusion of planning desire. That project has undoubtedly become the world's most studied and discussed dam—a point that certainly does not mean the research is exhaustive and the issues are settled (quite the contrary).

Their environmental manager put it like this, once again calling attention to the materiality of a tactical engagement: "The company has to rewrite the social and environmental plans all the time, and they grow by three kilos every time they're rewritten. So people like me are called to Manila, the Asian Development Bank headquarters, to explain what's been done. I'm constantly making matrices and revising them, updating them to keep track of the work that has been done. The bank keeps demanding that the company take into consideration new issues raised by the activists."

In this case, institution hacking turned into a kind of *institution jamming* in which institutional processes shut down for their inability to simplify their environment on their usual terms. "These guys are never going to be on board. The Bank responds with their typical knee-jerk reaction to everything that comes their way, everything. These guys, the activists, are copying everything to the president of the bank—it's the electronic revolution, right?—and they're all the time meeting with key people in ADB who are the decision makers. And there should be an appropriate way to actually filter this email but there's just not.... The real problem now is the Bank's inability to manage their own process," the manager complained.

A senior manager of that US$1.4 billion dollar project protested to me, when I asked him about IRN, that first they demanded extensive new studies of the social and environmental impacts, even after two sets of studies had been completed. "And now," he said, exasperated, "they accuse us of confusing the issues with science!"

This slog of research and planning obligations hinge on the very presumption of a system of rationality in which those who are in charge profess responsibility for how things turn out. Yet that presumption reveals itself to be logically impossible when questions of detail, control, and predictability are applied to the complexities of society and ecology. In forcing the issue, for a project the World Bank had explicitly set out as a model for how to do sustainable hydropower, activists baited planners into creating their own reductio ad absurdum. What Donna Haraway (1988) calls the God Trick of singular vision becomes a severe liability when the deity's coherence is rigorously audited by tireless, crafty opportunists.

Daley, the manager, never really believed that the problems were all that serious. Ultimately he was forced to take on a wide range of environmental management burdens for which he was ill-equipped, and he left the company shortly after the battle with activists was lost. (What those new management burdens entailed is the subject of later chapters.) Sometimes implicitly, sometimes explicitly, the failure to manage and control the natural environment collapsed into the inability to manage the organizational context (environment) in which the articulation of authority makes sense. Furthermore, the sense of "environment" is precisely that of a tacit context, the taken-for-granted ability to move and act against a passive ground that plays backdrop to the intentionality of rational agents. Daley described this collapse of that tacit ground as a crescendo: "We always had environmental people. We thought everything was fine. But then obviously, a couple cracks appeared. You get a village complaining about this or something like that. You know, ok, fine, we have to look into this, look into that, look into the next thing. And it sort of built up to a crescendo. And that's fine, we decided to solve it." That crescendo was the sound of a forced admission that his company had become vulnerable in ways he failed to comprehend. Activists had subjected the hydropower company to a labyrinth of unknowns that was of its own making.

Conclusion

IRN's work made possible a unique sociotechnical configuration. IRN described its work as documenting issues of concern and raising them in high-profile settings. This description accurately portrays the spatial strategy of undermining the ways an institutional network insulates itself from environmental issues and goes part way to characterize the kind of diagnostic work involved. It probably overestimates the extent to which the campaigns merely broker or relay the concerns of local people. IRN is highly selective in how certain issues are made into matters of concern. Put differently, while there is much dedicated labor involved, their work is also opportunistic. It unabashedly emphasizes certain issues and ignores others, just as it takes advantage of certain material and institutional possibilities but not others. Their brand of institution hacking explicitly works to undermine the tacit conditions of hydropower planning and construction. It is problematizing and does not propose solutions or definitive answers. For this strategy to work, their entrepreneurial outlook must be attuned to the technical potential for flexibility, play, and opportunity within existing relationships.

Furthermore, the very stuff of critique for explicitly political actors is already at work in the practices and condition of their vocation. Like engineers and scientists, who are concerned in their own ways with the effectiveness and instrumentalism of their own activity, activists cultivate a materialism of tactical effectiveness, that is to say, a repertoire of sociotechnical practices explicitly concerned with strategic results. For transnational activists such as IRN, this repertoire does not work through mass protests of popular demonstrations or civil disobedience; nor does it work through a counterscience or legal strategy. Instead it undermines the technical details of the assemblage in question. Crucially, these details include all kinds of heterogeneous elements, from image and reputation to financing deals, from complacent attempts to sideline village ecologies to the expert evaluations of particular design options—anything that can help form a crack. By undermining the delicate infrastructures of an industry, IRN engaged in the systematic production of uncertainty. Such a mobile, diagnostic approach also works through open-ended translocal and transnational spatial arrays enabled by loose organization among disparate but politically allied groups.

The sustainability enclave that emerged surrounding this project is a direct outcome of foreign activists' attention. As I have shown, their work partly included transformative, long-term relations with locally based activist researchers, but it never took the risk of long-term local engagement with Lao environment and development practice. Certainly for this project, activists never attempted to forge a substantive dialogue with the Lao bureaucracy, preferring instead to target multilateral banks and transnational firms whose bureaucratic pathways provided adequate tactical potential. Simply put, transnational activism established the terms on which environmental problems have been addressed, and their relatively distanced engagement had led to a tenuous set of ecological obligations that nonetheless have important effects. As will become only increasingly apparent, Lao hydropower's form of concessionary rule delegated certain kinds of environmental rule to the sporadic and highly variable interests of different foreign actors, from activists to engineers. Sustainability and contestation require each other, and Laos' sustainability enclaves emerged where transnational activism took root.

I use the term "technical entrepreneur" to identify a commonality shared between activists and the sustainability managers, introduced in chapter 3, who met them head-to-head and played their game on the terms activists had defined.[11] Technical entrepreneurs are tacticians of specific relations who explore and manipulate their immanent possibilities. Whether through dedicated labor, improvisation, or opportunism, the work necessarily takes the form of detailed technical knowledge of the relations in question. The work is not based on an empiricist or "scientific" grasp of the state of affairs, but on the multiplication of unexpected relays and sutures within a hierarchical system of finance and development planning. Importantly, their work is inherently protensive, for it is concerned with the tactical reimagination of possibility. *What can these real relations make possible?* Activists repeatedly exploited the unexpected transversal possibilities of institutional relations, communications media, and material vulnerabilities. In doing so, they demonstrated the actual conditions through which dams are built and the openness of apparently closed bureaucracies to new forms of practice.

By taking advantage of the play of threat and opportunity, technical entrepreneurialism does not establish the authority of its version of the facts. Instead, it problematizes and undermines. It probes and exploits. It establishes novel connections and explores what achievements might be possible. It uses the capacity of a relation against itself, thereby undermining its premises or confirming its potentiality—but either way, performing the potential of the relation itself. Is there any better demonstration of the truth of a relation than to show that it is subject to "unexpected reversals" or, in Latour's (1999, 281) apt phrase, "the surprise of the action"? Through their dedicated opportunism, activists demonstrated what the hydropower industry is by showing what it is capable of doing, just as hydropower developers have shown what a river is capable of by building dams. Activists maximized the potential of a bevy of specific relays or techniques, from empirical field research to email, diplomatic networking, and detailed technical studies, fleshing out the aporia, the denials, and the vulnerabilities of a privileged, protected industry.

On the other hand, by showing what these developers, banks, and dams are capable of, the events in question performed the being of the hydropower developers themselves, provided we recognize that *who they are* is hardly permanent or fixed. In this performance it became possible to identify the willingness of industry actors to base social and environmental decisions on conjecture, to bully villagers, to quibble over paltry sums of compensation, and to ignore the essential role of fish in the culture and the nutrition of the people whose lives are intimately tied to these transforming rivers. Only on this performance was it possible to identify the intimacy between financial institutions and developers, and the limits of that intimacy.

Could we name such a river as this—a river in which sometimes much of its water intermittently shifts to an adjacent river basin, to a river channel that never in its geological history was required to support such volumes? In the industry, it is called a *transbasin diversion scheme*—yet that describes not the river but the machine that performs its novel articulation. What forms of livability can such a river entrain? What could be the geohistory of an anthropogenic river? If Lao rivers have been the locus of desire for diverse, cosmopolitan attention, then this certainly holds true for the activists and activist-scholars who find themselves obligated, perhaps in unknown ways, to the demands called to them by these odd rivers. There is always the possibility for more. A hydropower configuration such as this is unlikely, and in its enactment it demonstrates not the epistemic veracity of the knowledges that created it, but simply its own ontological possibility.

The events, tactics, and techniques I have described can and should be interpreted in terms of a contestatory materialist politics concerned with its own effectiveness. But what I have described is also the very substance of living with hydropower, in which the intimacy of a series of connections brings together people and rivers in an uneasy collectivity. Who these people have become is a function of their involvement with a river transformed. Emphasizing contestation tends to downplay the extent to which political struggle conforms to ecological time and the obligations of acting. As this ethnography unfolds, the work of activists, hydropower developers, and multilateral finance banks slowly helps to demonstrate an inevitably partial answer to the difficult question *what is a river?*. In fact we do not know what a river is because the question is temporal: *when* more than *what*. For it is the differential temporality of a river, its constitutive underdetermination, that introduces decisive possibilities for living, including those that are dangerous and disabling.

Anthropology of the environment is liable to make a strict distinction between the river itself—its mud banks, its riparian flows, its immediate forms of life—and the kinds of sociotechnical practices described in this chapter. Email campaigns, concerted effort to crack open hydropower development institutions, or back-channel communications with bank staffers surely do not qualify as *ecology*, do they? What does it mean to live with rivers and, by extension, to live with hydropower dams? This question is at the core of activists' practices, not only because they ask this question of the people whose lives are intertwined with those riparian flows, but also because they materialize in their own practices the transnational ecologies through which riparian lives enter into a labyrinth of unknown power relations. These power relations manipulate and mutate the physical and biological relationships that previously enjoyed a different autonomy, and they introduce all sorts of heterogeneous elements into the ecological assemblage. Financing and constructing large dams are paradigmatic ecological activities. Likewise, the forms of privatized indirect governance affected by IRN are inherently ecological because they deepen an ecological problematic and have ecological consequences. By the same token, it is only through their engagement with anthropogenic rivers that these activists and managers' practices came to take their actualized form. The sutures and relays connecting rivers, institutions, and modes of activism work to make people conform to rivers differently.

How is a river ontologically distinct in relation to different groups of people or at diverse moments in time? In chapter 1, I demonstrated that a swarm of cosmopolitan activity can and does emerge around the anticipation and promise of future dams to come. The dams temporalize through the imagination of diverse futures. That imagination of environment—that fantasy of power predicated on a future of rigorous control over the gravitational flow of water—creates an obligation to rivers through which engineers and developers conform themselves to the specific requirements of their practice. Every developer can speak to the intricate, intimate details of failed projects that haunt his or her risky, capital-intensive wager against topology and hydrology.[12] One is inclined to think that once the dam is built, then it is established in fact and it is no longer a question of imagination. Yet the swarm of potential never ceases. The Theun and Hinboun Rivers, now conjoined through a technical machine, remain an open question. Just as the hydrological machine turned two rivers into a strange amalgamation, the work of activism provoked a retrospective evaluation, an affirmation of rivers and life conjoined in the interstices of networked power relations, the chains of email correspondence, and the anthropogenic capacity of a distinctive political modality. Michelle Murphy argues, "Latency in ecological time describes how the submerged sediments of the past arrive in the present to disrupt the reproduction of the same" (2013, 106). This retrospective imagination of environment holds open the alterity of the dam itself, maintaining its potential against a past that cannot be understood immediately and that resists the stabilizing discourses of technological triumph. Given the right tactics, hydropower developers can be the subjects of latency no less than affected villagers.

Looking to chapter 3, what could it possibly mean to "solve" the problems caused by the dam? The casualness with which managers and engineers dismiss severe and intractable problems, as if they could be solved and done with, is a surface effect, beneath which flows an abiding poverty of practical ability. Gesturing toward plans and budgets only works if someone can be hired who actually knows how to fix the problems at hand. And yet the evidence continues to show, in spite of prominent attempts to develop best practices, that nobody knows how to mitigate the social and ecological impacts of large dams conclusively. There is no reservoir of accrued practical knowledge that can dispense with the problem once and for all. In fact, this dam is interesting in part because its managers made a well-funded and genuine attempt to deal with many of the problems it had caused. But managing these late industrial environments, with all the "handiness" the term suggests, hardly suggests the finality of a solution. The nature of management is neither purified of human involvement nor a simple object of rationality and control. Managing problems rather than solving them suggests an articulated sense of living with anthropogenic rivers.

Interlude

Intimacy (Vetting)

I first learned of Richard, and by extension the Theun-Hinboun Dam project, very early in my preliminary fieldwork, in 2002. Expatriate consultants who were long-term residents in Laos were an important first resource for reflecting on the scope of development practice for a country that had only liberalized its economy over the prior decade; they engaged in sympathetic rumination over what sort of useful research I might provide. Letting these mainly older, mainly male Anglophone experts contemplate a role for me was one way to insert myself into a network of people that formed a sort of connecting web between large foreign donor projects and the Lao government and project staff who formed the practical working context for doing development.

One of these consultants has been characterized in a brief piece by Michael Dwyer (2014b, 101), who offers a portrait of a tragic figure caught up in a political economy of ignorance. "The mitigation expert," Dwyer writes, by way of describing this person, "for better and for worse, knows this economy intimately." There is a way to read *Anthropogenic Rivers* as an ethnography of white men in Asia, and the attentive reader will have noted that subtext along the way. If I foreground intimacy and its relation to the limits of knowledge, it is because that seems appropriate to a characterization of this iteration of white masculinity. I grew up around this stuff; my father is an environmental consultant who had worked in Laos once upon a time, and he put me in touch with several of my first contacts.

So these guys had many ideas about what I might usefully research. Most of the ideas entailed me getting the inside scoop on how the "government really thinks" about some certain issue, or a variation, me helping reveal the web of personal relations through which "real decisions" were being made. Others thought I should look at longitudinal effects of older development projects—a kind of rural sociology of development effectiveness. What became clear was their commitment to Laos, their trust in development as a process, and their willingness to think openly about the complex conjunction of social, political, and technical features of project work and its limitations and modest successes. Many of them had been initiated into development work through the visionary globalism of high modernist belief in technocratic achievement, only to become seasoned disbelievers in state-orchestrated growth. One of the most common themes was the importance of long-term relationships to the country and the "art" through which some development projects achieved some modest successes. It was the man Dwyer writes about who pushed me to look more closely at Richard's experiment with private sector, entrepreneurial sustainability management.

Richard's role at THPC had come with a potentially powerful narrative that had caught several of these men's attention and suggested questions central to their hopes and dreams of national improvement. Another manager at the company had framed this narrative for me, reflexively making clear that it was a tale—"Everyone is good people," he said. "Mistakes were made but we have taken responsibility and everyone can be forgiven." Could a hydropower project reform?, these long-term expatriate development experts wanted to know. Given effective leadership and financial commitment, could the problems of a dam be dealt with? Conversely, was it just narrative? Was Richard *sincere*? Far from being easily answerable one way or another, the questions stood like open wounds among those committed to and yet betrayed by the promises of development. Richard's charismatic and controversial presence within this group exposed an essential ethnographic problem. What is the hope in which Richard's understanding of environment trafficked? If Richard was a seductive figure, I came to realize, it was because these men wanted to believe, against the evidence, that hydropower could be a force for good. In a reversal that still astounds me, I learned from their axis of desire that, for those whose dreams had been betrayed by hydropower, the risk was not that Richard was insincere but that he might be telling the truth.

This interlude is a story about Richard, who is a kind of metonym for performative sustainable hydropower, but to get the sense of it I want to contrast his relational practice with the authoritarian hydro-utopianism of James. I spent much of my fieldwork with James and his team of consultants, whom I introduced in the previous interlude, "What Is a Dam?" As mentioned, James was the only expert I encountered who insisted on himself as a scientist. Once I watched him give medical advice to a sick villager while insisting he was a doctor—even though he was a zoologist, not a medical doctor. I worked with him for a long time because, since he had worked on so many hydropower projects in Laos, I was convinced he was central the dynamics of how the industry functioned there. But he was not. On the contrary, his perspective was rigorously legalistic, authoritarian, and utopian, and his understanding of how the industry worked was similarly impoverished. (For their part, hydropower developers repeatedly hired his team because he was very inexpensive compared to big foreign consulting teams.) While he had interesting things to say about the process of doing environmental work, ultimately it was his many fabulous, failed utopian schemes that underscored for me the radical difference in perspective and ethos he represented. Let me characterize this difference to draw out by contrast Richard's entrepreneurial ethos in which knowledge functions not in terms of correct representations of reality but in terms of ambitious, opportunistic achievements. The premise of utopianism, after all, is that if society could be organized according to universal science, harmony would ensue.

I got to know James because of his reputation for writing extensive environmental impact reports. He had an idea that environmental research could be used to bind foreign power companies to village customary law. He wanted to leverage hydropower to build something he called civil society, based on minute scientific documentation of environmental conditions. One plan was no less than eighteen separate volumes of ecological and social studies and action plans meant to provide the facts capable of supporting the legalistic enterprise. We agreed to meet at a local expat restaurant in the center of town where, having arrived early, I waited with a cup of coffee. He arrived some minutes after the coffee and (after scoffing at the price I had paid) he put me on the back of an ancient scooter, coffee unfinished and rode me to his office nearby. The stripped-down work space was filled with scientific equipment, assorted specimens, stacks of reports, and massive rolls of maps.

His vision of environmental planning implied a total approach that articulated economic, legal, and institutional structures into a seamless but fantastic apparatus. In his thinking, hydropower could be an ally to local agrarian development, and the two could form a bulwark against the misrule of the rent-seeking Lao state. He lauded economic producers as the backbone of society and told me to read books like *Lords of Poverty: The Power, Prestige, and Corruption of the International Aid Business*, a popular antidevelopment neoliberal manifesto by Graham Hancock (1989). His idealistic vision sat uneasily with contemporary Lao hydro-power development, itself far less interested in the rationalism of an orderly society and the putatively sovereign subject meant to inhabit it.

The main elements of the different plans he crafted were a strict separation of environmental management from the Lao government, which was held to be incompetent and corrupt; detailed scientific understanding of ecological and social dynamics; a comprehensive, albeit schematic, enumeration of mitigation activities, costed and complete with a projection of ten-year timetables; and, overall, an emphasis on durable engineering structures and legalistic institutional arrangements with villagers. A maverick gesture toward total planning, his approach to environmental knowledge displayed the utopian dream that correct understanding of reality could form the basis for its ideal political ordering.

There is no question that James's approach was far more labor intensive than what THPC would have been prepared to initiate, and much of that labor revolved around the scientific work meant to map socioecological complexity as the basis for comprehensive planning. (I have already mentioned James's Borgesian attempt to develop a spreadsheet calculator of the total environmental costs of hydropower impacts from the ground up.) But in practice this took the form of a fundamental pathos regarding the inadequacy of reliable knowledge. Notwithstanding the sheer improbability of writing a ten-year plan that could predict in advance every environmental management activity that needed to be undertaken, every aspect of his plans relied on accurate knowledge and comprehensive procedure, the seamless articulation of truth and power.

The inadequacy of method to truth claims resonated, in his writing, in the pathos of knowledge as the basis of action. Consider this passage: "studies [are] needed to actually ***understand*** better, the changes which the project has induced" (emphasis in original). Within James's practice this resonant threat of skepticism drew on a latent apprehension of complexity and suspicion about veracity and empirical reference. If environment, like society, is taken to be infinitely complex depending on how closely one looks, then knowledge cannot help but be perpetually inadequate. The result is a heroic effort to control for details.

There was also a juridical element, which could be said to define his understanding of the subject. The legalistic impulse was displayed in his emphasis on representational documentation and guarantees of contractual agreement. James insisted that villagers should be given letters detailing their compensation packages so that they could then verify the work of public authorities. For resettlement, he argued that the plans for each village should be posted in written public notices rather than the usual verbal communication, open to miscommunication, deliberate falsification, and error. While working on one project, he even insisted the project manager communicate with him and his team in written form because he felt the project directors tended to waste his time, misconstrue details, and deny what had previously been communicated.

The legalistic impulse was all the more apparent in the methodological problem presented by the figure of the villager. I have an archive of recordings his team made in which villagers were asked to state on record that they understand that they are perfectly free to say whatever opinions they have about the dam, and that they know that whatever they say will not be held against them. This archive is rigorously organized with the recordings (most about thirty seconds long) cataloged by the speakers' names, their official titles (for village heads), date, and village name. From the written report: "The interviewers . . . were well known by all household members, and seen to be the sort of people with whom villagers could be frank. Each taped household interview began with a question asking whether they felt free to answer the questions and whether they were worried about saying 'bad things' about the project. Householders signed to say they would tell the truth and say what they really felt."[1] What obsessions of rational subjectivity underwrite such a fantasy of obligatory free discourse?

The approach becomes a perverse performance of its own logical imperative, unable to be propped up on its own authority. James's desperate efforts at self-reference—marking the fairness of his own plan and his able rapport with villagers—simply indicate the auto-deconstruction of his overweening commitment to scientific truth and reason. The result is a kind of mad despotism. ("One thing at least is certain," Michel Foucault once wrote. "Water and madness have long been linked in the dreams of European man" [1988, 12].)

That digression hopefully fleshes out some of the ethnographic dynamism of the industry and underscores what I contend is a markedly different modality and practice of reason that characterized Lao hydropower development more broadly. By contrast, the practices of sustainability management I describe in what follows were highly relational (for they were predicated on negotiation and constructive encounter) and depended on a view of subjects as sociocultural beings susceptible to productive manipulation rather than abstract juridical subjects. The contrast in these two figures could not be stronger between authoritative knowledge ("what action is required by logic and fact?") and achievement-oriented practice ("what might specific, real relationships make possible?"—"what might we pull off," or differently, "what can we get away with?"). The latter is a subject position from which one can do whatever one wishes as long as it can be gotten away with.

Richard's collaboration with the activists, which I take up as an experimental form in chapters 3 and 4, was highly controversial within the hydropower industry, and in fact he was seriously reprimanded by his company's board of directors in an event that I was implicated in. Richard had simply failed to vet his experiment with IRN with his board, and at a certain moment, a rumor spread that he was bringing activists to the dam site to let them observe the environmental work. That was inaccurate, but I perhaps had started the rumor inadvertently. For earlier that week I had met up with a number of hydropower consultants, including James, and I had mentioned Richard's experiment with the activists. We talked about it a bit and people were definitely intrigued that Richard would do such a thing. James, who had no love for Richard, was also personally close to one of the board members, but perhaps it is just a coincidence that a few days later Richard had been severely chastised by his board and forced to abrogate part of the experiment.

I thus come back to the problematic posed at the beginning of this interlude. What kinds of play, promises, or pleasures are trafficked in sustainability management? Richard was in the position of carefully vetting his company's environmental work with a group of foreign activists, and yet he had wholly failed to vet that risky undertaking with his board. *Vet*, the verb form, arose in the late nineteenth century and referred specifically to medical examination of an animal or person for good health; the current usage, to "examine carefully and critically for deficiencies or errors,"[2] can apply to evaluating ideas as well as people to gauge their suitability or potential failings. And is not vetting an image of intimacy or, at the very least, flexible boundaries between ostensibly opposing political groupings? Why would Richard go out of his way to make sure activists had a chance to comment on and critically evaluate his proposal while at the same time hiding its controversial elements from company oversight? Rendered in terms of personages or individual figures, James and Richard represent very different masculinities at work—one predicated on authoritative control and the other on negotiation and open-ended manipulation of possibilities; one a figure of suspicion and juridical requirement, the other a figure of a certain pleasure in risk. Cohen (2008, 35) describes this pleasure in risk as "intermediate charisma," that is, the pleasure of mediation as partly an end in itself. If nothing else, the risky endeavor confirms that the space of anthropogenic emergence, created in part by activists and the river itself, was open-ended and subject to possible experimental formations.

Chapter 3

Performance-Based Management

March 22, 2004. I arrived at 7:15 a.m. to wait outside the main offices of the Theun-Hinboun Power Company. I stood across the street with a small backpack, trying not to look conspicuous. A team of external review consultants was in town to conduct a performance evaluation of the company's environmental program. Richard, one of THPC's managers, had promised that I could go to the dam site with them, but I had not been able to confirm any details except for the date. Fearing the promise was empty, I had decided to show up to see what might happen. Morning traffic swirled in front of the company's chrome and bright blue corporate logo. I loitered behind several cars, watching to see whether the company van would appear before making my entry.

The performance evaluation was part of an unexpected collaboration Richard had forged with IRN, the activist NGO. Richard had gone out of his way to include IRN in the process of designing and implementing an external review of their environmental work, in an effort to characterize their effort over the prior four years with some legitimacy outside of the narrow world of hydropower development and consulting. IRN had helped select the consultants, had given feedback on the terms of reference, and would be allowed to comment on the draft report before the consultants finalized it; this kind of outside scrutiny was and is unheard of in the industry, and the experiment was striking for its boldness. Whatever one might think about the trustworthiness of prominent energy company, the fact that Richard had already delivered on the first two meant this was clearly not a facile ploy. The joint audit was already three weeks into a four-week consultancy. The company certainly had a stake in being a good corporate citizen, whatever that might mean in practice.

After an hour or so, with the consultants and their van nowhere in sight, I decided to go inside. The receptionist confirmed that the van was already en route and asked me to wait. She offered to call the van back to pick me up, then balked; we waited for Richard to finish a phone call. Nine o'clock approached. Richard emerged in his characteristically brisk and friendly way—*so American*, I always could not help but think; crisp white business shirt and khaki pants, disarming informality mixed with decisiveness. "Sure, you can go down to the field site, as long as there is room in the van and the guest house." He confirmed these things, had me sign a release and himself signed an approval for expenses. The van was called back. "It's fine," he said. "They needed to come back to give me their signed contracts anyway." Strange, I thought. Later I learned it was only a coincidence that they had started work without signing the agreement. While waiting, we chatted about their expansion project, how much easier it would be with private financing than using financing with public oversight. The company's board of directors was in town for signing the most important stage of agreements that would allow construction of an additional dam to begin in earnest. Concerning the joint evaluation I was here to observe, he casually mentioned that IRN would not be able to look at the draft external review report after all—they would need to wait for the finalized public report along with everyone else.

As soon as the three consultants arrived it was clear something was amiss. We stood in the front foyer of the office, amid natural lighting and a tasteful rock garden. One of the consultants moved off, like the matters at hand did not concern him. Richard asked the other two for their contracts; they replied that, actually, they needed to talk about the changes—and they wanted two more days to think it through. It turned out that Richard had been obliged to go back on his agreement with IRN by cancelling the third and final part of it. This had meant changing the consultants' terms of reference by removing a symbol and an acronym—"& IRN"—from a critical passage.[1] Two consultants, Steve and Umporn, were threatening to quit, potentially leaving Richard with a high-profile public relations disaster. I stood awkwardly among the office plants, afraid to take notes lest I be asked to leave, while the social drama unfolded and Richard displayed the full power of his managerial acumen.

I listened intently while Richard expressed his own disappointment and attempted to convince Umporn, a social specialist, and Steve, a fisheries specialist, to finish their work. One line of negotiation was conciliatory. Richard expressed how disappointed he was at this turn of events and accepted responsibility for not having properly vetted the experimental joint evaluation with his board of directors. Mistakes happen, he said; if anyone is to blame it was him. The directors knew about the main elements of the joint evaluation, but not the details, he said. The play of events slowly became clear as he asked me whether I knew anything about a rumor that he was bringing IRN to his dam site to investigate environmental impacts. It was only a coincidence that his board of directors was in town; if the joint evaluation happened three months ago it would not have turned out like this. People in the industry, he remarked, have a simplistic way of negotiating with NGOs—for them it is just, "No, no, no no!" Next time he would have to rethink his approach for how to accept input. This time, there was little chance the board would go back on its firm directive to cut IRN out of the loop. He had talked to IRN already that morning and wanted to finalize the change in plans by that afternoon. He took a rhetorical position in the middle ground between THPC's board, which could not see things his way, and the consultants with their NGO proclivities.

An important element to his performance was hard for me to identify exactly, but it clearly related to Richard's ability to span old political divisions and reach out to diverse constituencies. It was not only that he had forged a working relation with a highly antagonistic activist group; neither was it limited to the appeal or optimism he offered Vientiane's development circles. He communicated a particular savoir faire of embodied finesse that communicated his reliability beyond the content of his speech. As one of the California-based activists put it to me a few weeks later, in spite of intense debate at IRN whether to work with them, "we actually trusted Richard."

Furthermore, his performance was not one of authoritative expertise; on the contrary, he performed relationships; he performed an orientation ("this is what we're going to do"), joint effort ("we're in it together"), and thus a certain kind of promissory inscribed onto the body of his person. Richard articulated commitment as a kind of collective project. He performed the weight of IRN's activist campaign—validating their concerns in his own partial way—through his personal commitment to negotiating some sort of resolution. Richard's intimacy was a promise; uncertainty management could be read onto his body, in gesture and verse.

Two days later, we were in a van headed to the dam site; Steve and Umporn had decided to go ahead with their work. But in the meantime, I had discovered an unexpected problematic. What could it mean to manage uncertainty? What is a manager—what is the content of that labor? From whence derive its intimacy and willingness to banter and hold forth, which are so foreign to the engineer? And, if damaged rivers get managed, if pests or erosion or nuclear waste gets managed—if "management" is a paradigmatic activity for late industrial environments—what exactly does that practical activity entail? How did Richard perform that slippage between managing the problems caused by the dam and managing the people who were able to protest those problems? For it struck me that far from denying the problems caused by this dam, Richard had formulated a manner of acknowledging them and, therefore, staged a certain kind of affirmation of anthropogenic rivers.

Living with Anthropogenic Rivers

In this chapter I put forward the argument that management is a particular kind of practical reason that bears directly on late industrial environments. If we take neoliberalism as a program of "cheap government" (Rose 1999) that delegates many problems of government to domains of market practice, then one of the problems to be considered is that businesses are often de facto in the position of contending with any number of issues tangential to what they consider their core economic activity. It seems reasonable to suppose, therefore, that business knowledge deploys its own vernacular materialism and related modalities of skill and expertise when it comes to managing socioecological relations.

I propose a stronger claim, namely that management is the hegemonic form of living with late industrial environments. Anna Tsing asks for an anthropology that can "stimulate a vocabulary for livable disturbance—a first step in coming to terms with the anthropogenic environment our species has created" (2015, 93). "Given the realities of disturbances we do not like, how shall we live" (92)? I here take up management to pursue a pragmatics of anthropogenic ecologies from within the framework of capitalist enterprise itself.

In particular, management is worth paying attention to in part for the ways it affirms, grapples with, and sometimes deliberately produces ecological uncertainty. Whereas the production of uncertainty was an outcome of activists' practices, environmental management in this case turned the production of uncertainty into an outright practice within a genealogy of similar management techniques. While the production of uncertainty is sometimes deliberate, it always occurs within a history of practical techniques ranging from the development of risk communication procedures to the protocols for managing industrial chemical accidents. On the one hand, management thus betrays a comfort and familiarity with uncertainty that is an important part of its ethos, while on the other hand, it implies a whole body of ecological techniques and practical dispositions that cannot be described in terms of rational control or mere destruction (let alone profit seeking). This middle ground of techniques, uninterested in either the purification of a nonhuman nature or totalizing mastery and control, constitutes the dominant mode of living with anthropogenic ecologies today, and its relationship to uncertainty is central to properly understanding it.

Hailing from Annapolis, Maryland, Richard came to Laos in the mid-1990s as an attorney for Laos's first registered law firm. He also had a background in mechanical engineering, including an engineering degree and a diploma from General Electric's nuclear power training program. In the context of his working relations, I understood that he came to the company with experience in human rights law and nuclear safety. He had been encouraged to apply for his position, in spite of having no previous experience in hydropower, by the company's earlier general manager, Jay Daley. As Daley put it, "He was applying for the job. He gave me his application, I looked at it and I said 'Richard, this is crap.' He was coming across as a lawyer who had engineering experience somewhere along the line. [I told him], 'No, no, you're an engineer who has done some law work.'" He approached his managerial dilemmas not as an engineer who has come up through the ranks and earned the skills to manage an entire company, but as a manager per se who was able to organize around him the means necessary to keep things running. What he brought to the company was an outsider's perspective, familiarity with industrial risk management from his nuclear background, and a lawyer's sense of discursive flair. While at times his professional credibility suffered, his outsider status established critical symbolic capital with people outside the hydropower industry.

He described his approach to sustainable hydropower in a way that, in broad terms, was largely borne out in practice, although it neglected less favorable aspects of his work. In one respect, he argued that environment should be treated with equal importance to the company's commercial operations, rather than being delegated to a remote office or a third party where it would not receive its deserved attention. Second, environment needed to be addressed with a combination of good ideas and skills, good management support, and proper financing. In another respect, he said that his strategy with IRN was to get them to change their approach of blanket criticism of dams by engaging them in the collaborative joint evaluation. I take the claim seriously, for it seems clear that his objective was partly to draw IRN into a compromised position in which they would have to give a positive evaluation of THPC's work. It was also apparent that he viewed the risky endeavor as a potential feather in his cap that would augment his personal career ambition. (At one point, he was mocked by other hydropower colleagues for bragging it would earn him a position at the World Bank.) But his approach also put him in the compromised position of working with IRN, regularly vetting certain ideas with activists, and paying fairly close attention to what activists were saying.

Lastly, he put in place an approach to environment that he called using "development tools without development rules." He along with an environmental manager, a New Zealander, were skeptical that extensive research should form the basis for environmental interventions, and they questioned the need for paying expensive international consultants when that money could go directly toward interventions. They emphasized a proactive approach involving a range of participatory mechanisms rather than simply trying to rectify specific problems or return village livelihoods to a prior condition, and they built the environmental program around a technique of stimulating entrepreneurial motivation among technical staff. "Development tools without development rules" was offered as a private sector approach to the frequent failures of public sector or NGO development work.

Richard's management practice was an instance in which industry actors were forced to think through the shifting obligations to the effects of his company's dam. His practice should not be understood only as a means of control or manipulation but as an attempt to rethink the kinds of environmental obligations hydropower should bear. "Obligation," writes Isabelle Stengers, "refers to the fact that a practice imposes upon its *participants* certain risks and challenges that create the value of their activity" (2010, 55, her emphasis).[2] How exactly could hydropower be done differently? The valences of managerial reason, its temporalities, and the world it helps compose are the questions to be explored; as Stengers argues, "Rationality . . . becomes synonymous with risk and challenge" (52–53). American managerialism was the immediate practice through which these novel obligations were put to work and evaluated.

More broadly, novel obligations to anthropogenic environments therefore come to bear on the grammar of the term "management." This is clear from assessing the objects or targets of management. If one does a keyword search for "management" or "to manage" within peer-reviewed journals, a wide range of objects are linked to practices generally described as managerial. These tend to be strikingly unique objects. Pain, diabetes and other intractable diseases, mental health conditions, traffic, sewage and storm water, risk and crises, forests, rivers, watersheds, logistics, organizations, information, industrial labor, disasters, chemical leaks, large crowds, pests, product design, and financial management—in all of these complex objects, contested norms of human existence are tightly linked to forms of practice and denatured natural processes.[3] Together perhaps they are not objects but problems—"problems for management." They are *problems* as sites of ongoing, systematic, and potentially innovative work on long-term, intractable relations involving a certain intensive labor commitment. And they are problems *for management* in the sense that they have been delegated to management and even actively reorganized so that managers can handle them. These problems are not solved; they are, precisely, managed. Management is one dominant, hegemonic modality of living with anthropogenic environments, for it variously constitutes techniques for the reinvention of nature, a biopolitics of letting life die and making life live, and the materialization of novel ecologies.

Image, Affect, Network

There are many valences to the verb "manage," a word with the Latin root *manus* (hand), which has come to be used in contemporary English in all sorts of contexts. Salient for my purposes is the sense of the activity of "getting along" or even "managing (barely) to get by"; it conveys the radically different sense of the activity of management from the term it replaced, namely "administration," which presumes rationalization and bureaucratic control. One cannot administer metal fatigue or diabetes, but both those objects are natural targets of management. Furthermore, if management seemed to be the paradigmatic organizational activity, it is also true that the term spread rapidly to a great many different kinds of objects, such as crowds or traffic. In fact it was adopted in English from a context closer to the shop floor than the office—the fifteenth-century Italian *maneggiare*, the labor of tending to horses, also dexterity and "artful trickery."[4] Consider the skill and risk of breaking horses or, to extend the equine metaphor, of harnessing rivers. *To manage* suggests the work of keeping intractable, heterogeneous, and sometimes unknown processes from exceeding critical thresholds. It suggests a working relation with forces that have their own autonomy and momentum and voids the prejudice that problems are met with seamless solutions. If practical reason is always a question of "what to do when," then, at least in part, management is what is done when things threaten to get out of hand. Management's materialism is the answer to Ulrich Beck's (1999, 18) question, "How will we handle nature *after* it ends?"

In the wake of IRN's activist campaign, Richard's role at THPC involved a deft diagnosis of the predicament in which the company found itself. Its major backer, the ADB, had abandoned the company to its own devices; the legal protections afforded by the Lao state were useless in a war over reputation—and THPC certainly could not have expected the government to actually fix the problems caused by the dam. To make matters worse, the developers had never collected any baseline data describing the state of affairs before the dam was built. No one knew how poor people had been, what their nutrition was like, or even how productive their crops and fisheries had been. At what point could the company say it had fulfilled its obligations? Could these problems even be fixed? There was another angle to Richard's diagnosis as well. ADB and Daley had been particularly ham-fisted when it came to working with activists. With no sense of nuance, they had simply botched the management of a delicate situation that had required care in communicating risks and expectations. In no small way, Richard's performative embodiment taught me to look more closely at the dimensions of image, affect, and network that characterized the activist campaign. Richard's person became central to managing the company's environmental problems, for it was his personability that allowed him to cultivate a direct working relationship with activists in terms of image, affect, and network.

Network. Richard's outgoing style served to allow activists to work directly with him instead of through the ADB's bureaucracy, thus reconfiguring the normative network they had built. He engaged activists directly, effectively working to eliminate activists' dialogue with ADB and redirect it toward his own person. Richard apparently sensed that the bureaucrats and public officials at ADB were not sophisticated in interacting with activists; he was willing to wager he would be better suited to cultivate those relationships. His practice effectively divided management of environmental issues in the field from the management of activists. This mimicked the operational distinction between risk management and risk perception management, that is, it redistributed uncertainty by construing the challenge of dealing with activists as operationally separate from that of dealing with matters along the rivers in central Laos. It was also based on the fact that ADB's bureaucracy could potentially enmesh the private company in a weighty game of passing the blame and adhering to stifling oversight. The firm's relation with ADB itself had become an uncertain variable, as THPC was essentially asked to absorb the cost and take the blame for the environmental problems when ADB had played a central role in making them.

Image. Richard subsequently sought to renarrativize the events surrounding the dam's impacts in order to rework the company's image. His innovation was to construct a multifaceted narrative of the company as reformed from past mistakes in order to shore up the kinds of rumors, images, and narratives circulating about his company. This was most aptly demonstrated by his actual cooperation with activists. Perhaps the single part of Richard's strategy most risky for him and for activists, the joint performance evaluation was built on the idea that an impasse had emerged in which activists would make a series of claims about the dam's environmental problems only to have the report rejected by the company, which would then make a series of counterclaims based on radically different assumptions about the legitimacy of large dams. In the absence of shared environmental norms, the joint evaluation was an attempt to produce a politically neutral epistemic base to evaluate the company's reformed and broadly expanded environmental efforts. Even as the joint performance evaluation threatened to unravel, the fact that Richard had risked it strikingly folded the activist group into the company's image repertoire. Controversy and dissent was real and alive, and yet part of Richard's image repertoire itself. "We're trying to convert them," said John, a New Zealander who set up the environmental management program, referring to IRN. "We want them to acknowledge that mitigation can be done well."

Affect. Finally, whereas IRN's campaign stimulated a rancorous, angry conflict over the firm's many liabilities, Richard's good-natured approach shifted the affect of the firm's relations away from animosity toward something characterized by goodwill. Indeed, there was a reassuring pleasure, tinged with unease, in his dialogue and personal interaction. This was particularly important vis-à-vis development professionals working in Vientiane, but he also went out of his way to give presentations to the ADB and in other venues where it made sense to assuage concerns and repair damaged relations. Disarmingly personable, he liked meeting and talking with people and used public forums as well as chance encounters in Vientiane's small expatriate scene to introduce himself, to ask questions of people who might be skeptical of his company's environmental work, and generally to maintain actively a sense of how people perceived his company. His informal interactions and his willingness to listen and talk about people's concerns were assuring and yet—precisely for this reason—disconcerting. People within the development community looked to me to figure out whether they were *really* fixing their problems or whether it was *simply* image management. "Richard's a lawyer," one consultant reminded me. "He's good at playing games." Perception management takes work, but even when done well the labor of personability cannot help but contain within it a false note of manipulation.

In terms of image, affect, and network, Richard's practice thus operationalized comparatively superficial aspects of activists' campaign at the same time it drew into relief their techniques. Much of his charisma hinged on developing arenas of informal interactions or casual relations. Richard's personable attitude worked to keep relationships informal and nontechnical by maximizing and localizing discursive interactions, many of them verbal and face-to-face, thus minimizing angry emails sent back and forth and disrupting the network practice that could amplify negative affect. Richard's wager formed one industry response to strategies of NGO regulatory power developed by activists, wholly distinct from the reductio ad absurdum in which other companies had gotten mired. This strategy of personalizing otherwise critical relations positioned him as the mimetic center of a network of relations struggling over the evaluation of environmental effects of the Theun-Hinboun Dam. He at once personified the politics surrounding his company and allowed that politics to come to bear on his personal demeanor and entrepreneurial ethos. Yet the persistent informality he cultivated could easily be seen in a more insidious light: he simply cut the ADB out of the picture, thereby displacing the activists' main source of real leverage.

No surprise that he cultivated a specific embodied habitus that displayed openness and a frank willingness to talk about what mattered to people. People familiar with the project but wary of the damage it had caused often remarked more on his specific role rather than the whole array of changes that preceded his arrival. Even after the falling out during the joint performance evaluation, one activist said they would consider working with him again even though they made the mistake of trusting him too much the first time. The country director of one of the largest development NGOs working in Laos said that he was the only private sector person that understood the needs of NGOs in Laos. Two of the three consultants who worked on the joint audit of the firm's environmental performance cited the innovativeness of the process as the main reason they took the job, and they generally praised Richard for taking the initiative even after it nearly fell apart. Once, at a public consultation luncheon for a World Bank project, while I was sitting with a friend and a handful of development NGO staff working in Vientiane, Richard came to sit with us—joking that he had come sit at the liberal table. I was surprised that he was not sitting with guests from other power companies or from the development banks, but instead he used the opportunity to talk with us about social and environmental issues. Due to his consistent labor of cultivating certain kinds of relations, he had clearly held out a persuasive image on the social development front.

A key element was his clearly articulated position about environmental impacts: dams have problems, but they provide important benefits and these problems can be fixed with the right combination of good ideas, efficient management, and proper funding. Contrary to the appearance of hiding his politics, his politics appeared as a central element of his self-presentation. This was notable for a context in which privatized hydropower development did not have an especially good reputation. Development consultants with strong social and environmental concerns consistently noted positive impressions of his strategy and welcomed the opportunity to learn more about his company's work. Many of these people envisioned large dams as revenue drivers that could underwrite much needed social development projects—except for the fact that they came along with such far-reaching consequences. Richard leveraged the possibility that the impacts of a dam could be minimized and the side effects properly managed. If Richard could get IRN—of all possible environmental groups—to participate in a relatively trustworthy report claiming that they had done a pretty good job, it would be a coup for the hydropower industry and a point of distinction on his personal vitae. For those who wanted to trust hydropower as a real development option, Richard trafficked in a certain kind of hope—a renewal of the developmental dreams of large dams. Their risk was not that he might be insincere but that he might be telling the truth.

Richard was therefore just as much a technical entrepreneur as the activists, but the technicism was para-ethnographic and interpersonal and involved a sensitive diagnosis of the predicament his company was in (on para-ethnographic knowledge, see Holmes and Marcus 2005). More than simply barely managing to survive the negative attention, he appropriated the potential of an unexpected array of relations to rework the context in which hydropower makes sense. By centering his own person as a node of public attention, he bypassed the clumsy public relations management of the public bureaucracy in favor of a multiplicity of informal relations. By articulating a narrative of reform, he rhetorically positioned himself as a mediator who could promise a different kind of dam that did not shirk its duties. And by playing on the faded hopes for large hydropower, he toyed not only with the allegiance of an important group of people but also affirmed the political rationale for a questionable development strategy. Kim Fortun (2001) argues that the stakeholder management form has an insidious quality that anthropologists have not fully engaged with, for the fact that it relies on the kind of situated, contextual knowledge that anthropologists themselves produce. Like activists, Richard was an entrepreneurial technician of the relation. As I argued in chapter 2, every politics concerned with its own effectiveness implies an ontology as a theory of its power. Image, affect, and network demonstrate that Richard's ontology was deeply social and relational, and his technical entrepreneurialism was para-ethnographic for the way it opportunistically redeployed existing relations of practice against a horizon of threat and opportunity.

Performance-Based Management

It is a mistake to attribute too much power to Richard as an individual, and there can be no anthropology per se of a person separate from the inherently social repertoire of habits and behaviors that constitute his or her individuality. Richard was a performer in both senses of the word—he acted, whether theatrically or otherwise, through the use of an existing grammar of possibilities, and in those performances he placed himself in the position of performing well or poorly, in contexts in which skill matters (on performance, see Butler 1993, 1990; cf. Mol 2002). Thus, my focus on Richard's person should not be read as a straightforward ascription of agency, for the instruments of his powers are not his, even when they so intimately invest his body (see Law 1997).

Yet the emphasis on managers and management as a structured domain of social practice goes farther than this. On the one hand, management provides a partial but extensive diagram of late industrial capitalism from the unexpected vantage of its problems and predicaments. In other words, what kind of problem does ecology pose to actually existing capitalism, and through what definite modes of entanglement do capitalist practices accommodate themselves to the sticky, real relations of specific ecologies? How, then, is capitalist practice a mode of ecological opportunism that produces novel ecologies and new forms of anthropos? On the other hand, management involves a definite array of techniques and practices that have a history. This history describes a long-term process of developing articulated ways of grappling with highly contingent situations in which all kinds of heterogeneous relationships and assemblages produce unexpected effects. The definite body of techniques encapsulated by the term "management" is in fact a partial and systematic diagram of contemporary socionatural relations; concomitantly, the person of the manager embodies diffuse sets of skills, techniques, and gimmicks necessary for the operation and maintenance of capitalism.

It is no surprise, then, that the vast body of management literature occupies a genre somewhere between self-help and how-to instructionals. With two references that came up during my fieldwork, here I provide critical touchstones for contextualizing the practices of Richard and his management team within a genealogy of American management practice. The first is perhaps not so substantial, but it provided an immediate, although partial grounding for the questions I had come to ponder—namely, what is a manager, and is this category of labor significant? What does a manager do?

I had found in an NGO archive in Vientiane a photocopied article, the title of which promised a partial answer to my questions: "What Effective General Managers Really Do," by John Kotter ([1982] 1999). Kotter's argument centered on the apparent disorganization of business leaders who do all kinds of different things throughout the course of the day, and whose hodgepodge of activity does not fit "into categories like planning, organizing, controlling, directing or staffing." But, he argued, "effective executives carry out their planning and organizing in just such a hit or miss way" (148). Based on daily time use surveys and tracking conversations throughout managers' days, Kotter provides an empirical touchstone for the popular literature on the habits of high-performing individuals. Performance is indeed the central point of validation. Among other things, it contained the apposite passage: "*'Excellent' performers* ask, encourage, cajole, praise, reward, demand, manipulate and generally motivate others with great skill in face-to-face situations" (153, my emphasis); performance and practice are linked through distinct modes of activity. Aside from the cult of leadership, Kotter's emphasis on the wide range of managers' performative tactics helped clarify what I experienced ethnographically with Richard.

Other elements fit nicely as well, and I wondered (but never learned) whether this kind of literature had been part of his self-formation. In particular, Kotter argues that many general managers ask hundreds of questions throughout the day; they seek out unexpected perspectives; and they talk about a wide range of interests unrelated to official business ("activities that even they regard a waste of time" [148]). Richard was practicing what I came to call performance-based management. (I even adopted Richard's pseudonym from a name in this article.) Performance, I discovered, was not only the performance of theatre or of gender theory; it must also to be considered as a native term internal to management's relationship to knowledge. Its central tenet is that *achievement is proof of concept*—or, more commonly, try things and see what works—and by correlation no matter how good your theory, if you cannot turn it into a successful enterprise then the theory does not count. In spite of the fact that Taylorism and Fordism have been paradigmatic for understanding capitalist labor relations, replete with images of determinism, positivism, and standardization, it was already clear from the ethnography that scientific management was far from the central issue at hand. The concern was rather with flux, variability, initiative, and personal dynamism under conditions of partial knowledge.

Another, perhaps more important reference emerged from the fieldwork to deepen this emphasis on performance. Already by the mid-twentieth century, Taylorist scientific management was being sidelined by the profoundly influential writings of Peter Drucker, who argued that management was not a science or a profession such as accounting or law, but rather a "practice" (Drucker 1954). I first learned of Peter Drucker's work from John, another manager of the THPC.

The central theme of the conversation was how to achieve results when lots of different factors are in play. My notes read:

"Peter Drucker—logical framework. Mgmt school for Amer. corp. practices. Taken up by USAID, how to integrate society, agronomy, physical, social—design serious activity get results. Now proper subject in [development studies]."

According to Drucker, management should be understood as a practice that undergirds the very idea of a free market enterprise. "The ultimate test of management is business performance," Drucker argued. "Achievement rather than knowledge remains, of necessity, both proof and aim" (1954, 9). Richard's invocation of the promissory, or an orientation of collective activity, seemed to luminously glow from these pages. "Anyone whose responsibility it is to act—rather than just to know—operates into the future" (Drucker 1954, 15). These passages capture an American pluralism, distrust of totalizing knowledge systems, and emphasis on results rather than foundations. Systematic empirical knowledge of the world in which one acts is explicitly sidelined in favor of a mode of action that is diagnostic and achievement-oriented.

For Drucker, an enterprise is not merely business; rather it is the organized, visionary engagement in larger-than-life projects requiring collective productive action. In other words, even with systematic automation, management involves the activity of orchestrating collective achievement. The case study with which he begins his most famous book, *The Practice of Management* (1954), is that of Sears and Roebuck's development of its turn-of-the-century mail-order distribution system. A couple observations are relevant here. First, for Drucker the innovative enterprise does not discover a preexisting market; it makes a market, where none previously existed, through the novel use of existing infrastructure. This was Sears & Co.'s famous mail-order system, which combined wide rural distribution of printed catalogs, a long-term strategy of winning over farmers' trust, a complex mail-order processing plant, and careful attention to high-quality products backed by a money-back return policy should anything go wrong. The creation of new markets is the essential outcome of capitalist enterprise, starkly contrasted with the work of the "shrewd speculator, buying up distress-merchandise and offering it, one batch at a time, through spectacular advertising" (29). For Drucker, enterprise forms an integral dimension of free society; society is free when people are able to act collectively toward an original achievement.

Second, for Drucker, management has a cybernetic quality that connects the expectations of customers and employees through the diverse communications media, physical objects, and laboring conditions that bring them together into a single, articulated formation. The quality of the physical commodity has to correspond to its characterization in the mail-order catalog, reenforced by matter-of-fact advertising that will appeal to a rural populace wary of "spectacular bargains" and "exaggerated boasts" (28). Correlatively, managing employees rarely involves telling people what to do; as with marketing, the manager must inspire. Customers are not outside the business—on the contrary, they are explicitly part of an anticipatory-communicative-physical mesh of contingent connections. The mesh (trap, snare) is to function as a coherent complex or even a material-semiotic web (although such language is foreign to Drucker) that concerns itself with the historically specific persons brought into the net of its organization. Far from any notion of autopoiesis, creating and maintaining this coherent complex is the specific task of managers.

Finally, the truly successful business transforms its own conditions of possibility. Enterprise must respond dynamically, not merely reactively, to the inevitable flux of social change; such a response implies a complete retheorization of the business model. In the Sears case, Drucker is concerned to show that by the mid-1920s the rural market was rapidly shifting toward small towns, and the centralized mail-order system had to be supplanted by setting up hundreds of stores across the American West. Once again we are shown the connection between strategic decision making and the material basis for reimagining possibilities, and Drucker's text articulates a vocabulary of midcentury materialist modernism: finding sources of supply and purchasing goods; the development of product design and manufacturers' capabilities; redesign of refrigerators, for example, for mass market; "systematic building of hundreds of small suppliers" for distributed production; creating the signature sales counter alongside new organizational structures and manager training programs (1954, 30–31). Management is the vernacular material practice that describes these compositionist enterprises.

Amid the text's heroic narrative and idealization of capitalism, Drucker is essentially concerned with achievement as the validation of enterprise. Any empirical assessment of market conditions or labor dynamics must be subordinated to a logic of action and decision, providing a stripped-down version of American pragmatists' axiom that real empirical questions must be motivated, in the long term, by concrete differences in practical courses of action (James 2008; esp. lectures 2 and 6). Profit is the outcome (and index) of the business model's premises, but these premises in no way can be reduced to a set of technical requirements, or to universal claims about the human and nature in the spirit of scientific management or modernist planning. (He later became disillusioned with the rise of investor-driven financial capitalism and turned his attention to the management of social good, specifically nonprofit organizations.) The performative excess of visionary ambition is an essential component of the entrepreneurial ethos, and this too bears a direct relation to the diverse kinds of relationships and objects on which management works. The ontology in play is not universalist or deterministic. Rather, diverse, real relations, many of them "merely symbolic," contain a potentiality that it is the technical entrepreneur's essential task to demonstrate—not by minute analysis but by performative achievement. The coup de grace of an achievement, not its conditions of possibility, is the enterprise's only validation; truth is a relationship to future potential or demonstrated results, not exhaustively analyzed preconditions. This ethos cultivates what might be called a pragmatics of underdetermination, that is to say, a repertoire of constantly pushing the limits of *what these definite, real relationships might make possible* or, put differently, *what we can get away with*.

These considerations invite two further observations. First, the ethos in question has an opportunistic relation to uncertainty that bears only an oblique similarity to risk as the statistical determination of probability

(Luhmann 1998; Hacking 1990). If risk is understood narrowly in terms of the quantitative decision making characterized by formal systems such as probability theory or social systems such as insurance, then threat and opportunity are essentially tactical and pragmatic rather than theoretical and formal. Their systematicity depends on the cultivation of case studies and repertoires of technique. Second, it follows that one cannot be an entrepreneur without cultivating an active and positive relation to threat and opportunity. The ethos views real relations as strictly underdetermined—that is, capable of being appropriated into new projects of achievement, or of throwing forward some unexpected contingency in the surprise of action. Thus it cultivates an integral, familiar, and productive relation to uncertainty as a mode of subjectivation or a relation to the self. In contrast, the very basis of risk politics has been held to be that social decisions are stymied when knowledge is inadequate to guarantee outcomes (see especially Callon, Lascoumes, and Barthe 2011). That understanding of risk, which comes out of European rationalism, ignores the seductiveness of risky, real relations. After all, most businesses fail and would never have been attempted without an economy of dreams (Miyazaki 2013) through which entrepreneurs dedicate themselves to improbable projects. Technical entrepreneurs are not masterminds but people who are seduced by extant possibilities and dedicate their persons toward forging improbable relationships. They are in a sense caught up in the pleasures of real possibilities. If there was a pleasure in Richard's full-body engagement with activists, it came from a certain subjected fascination with immanent possibility.

And yet such a modality is a recipe for a landscape of ruins. Late industrial environments are characterized by the production of uncertainty in part because this practical mode of ambitious achievement is inherently denaturalizing: it is predicated on undermining what is taken for granted and turning it toward some unexpected purpose. Capitalists produce risky environments through the cultivation of a specific ethos of risk-taking. Furthermore, if we take this ethos as an essential component of a "historical ontology of ourselves" (Foucault 1997c, 315), then it quickly becomes clear that it is part of such a historical ontology not because the ethos itself has become so generalized that it represents some broad swath of humanity. On the contrary, it is no doubt displayed only by a limited set of specific actors. However, it is through their activity that so much of earth's ecology has been transformed and hence defines what their contemporaries must suffer through.

Richard's practice bears this out. By initiating a risky and ambitious attempt to draw activists into a compromising, collaborative joint venture, he sought to rework the mess of obligations that had come to bear on the company's operations, a task that is not logically or practically inconsistent with actually trying to fix the problems caused by the dam. On the contrary, those problems are delegated to managers, who are given the dirty job of keeping at bay whatever contingent effects threaten to get out of hand—sometimes, even, by fixing them. For this reason, this kind of highly interested commercial knowledge, foreign to the discipline of economics, can be thought of as a subjugated knowledge of capitalism.

Subjugated Knowledge

Foucault describes subjugated knowledge as "blocks of historical knowledges" that were crucial to the work of programs such as psychiatry or military organization but that were "buried or masked in functional coherences or formal systematizations." Likewise, subjugated knowledges are those that are "disqualified . . . insufficiently elaborated knowledges: naïve knowledges, hierarchically inferior knowledges, knowledges that are below the required level of erudition or scientificity" (Foucault 2003, 7). The concept provided a way for Foucault to begin unravelling the long history of deep-seated materialism within Western thought and practice, in direct contradistinction with the theoretical reason that dominated philosophy (and political philosophy in particular), with its universal categories and deterministic baggage. The interest in practical knowledges was part of his departure from his concern with the discursive episteme toward understanding the historical underpinning of Western truth regimes, what he later sometimes referred to as a historical ontology. Subjugated knowledge is the pragmata of doctors and nurses, whose field of practice is the normalizing encounter of the clinic, rather than the theorizations of normal medicine. It is the knowledge of managers who must deal with all kinds of real people and situations, rather than that of economists with abstract models of autonomous rational subjects. Emphasizing this field of practical activity is an essential component of "reading Foucault as a materialist" (Murphy 2006, 181).

John was a New Zealander who set up the company's Environmental Management Division in response to activists' claims. He had been hired by

Daley after the company had been forced to take on new environmental obligations and had years of experience in commercial forestry in Thailand and working on a UN development project in Laos. I first met him in the company's headquarters in Vientiane in a meeting with Richard as I got to work on my research. About fifty, he had the rough appearance of a field hand—not without its own performative effect—and did not mind showing up in the office with soil on his boots or, once, wearing a broken pair of glasses held together with plastic twine and rubber bands. When I met him, he and Richard joked about two other researchers who had just visited the project, Scandinavian women who were "pretty nice to look at" and "not bad to sit across the table from for a few meetings." True to form, Richard treated these comments as a litmus test, watching my reaction expectantly to gauge my political sensibilities. (I demurred from responding.) It was John who brought Peter Drucker's ideas to the project, combining development experience of a sort with a private sector sense of getting the work done and not belaboring perfection—what Richard had branded "development tools without development rules."

My meetings with John were filled with many, many details about the environmental activities, including some meta-reflection on strategic issues but, as with Richard, very little personal narrative. With Richard's apparent openness, questioning, and dialogue about problems, I was left feeling as if a screen reflected back at me an image I had helped create. By contrast, John spoke a matter-of-fact grammar of objects and things, and his short, brisk sentences rarely rewarded a close reading. Yet his dialogue frequently had a certain irony about it. It was he who remarked, while gesturing inverted commas with his fingers, that the company has "a different narrative now—we're reformed." In another quip, he argued, "there are always lots of things happening here, but in the end nothing ever happens in Laos." This remark stuck with me. In context, the remark was meant to trivialize what he felt was overly dramatic concern over how certain environmental issues were playing out: people make too big a deal out of things, it suggested. Yet I came to realize it revealed a lot about his worldview, for the comment drew out a temporality of stasis. I learned from John that I had to watch carefully, for behind all the different activities, their profuse objects and apparent energy, was the structuring of stasis, an endless temporality of management by deferral. Managing problems—not fixing them—implies a kind of permanent, remedial abjection. Some tactics of this management by deferral were made explicit to me; others are imputed. Here are its key features.

Intervene rapidly in village environmental affairs to silence critics and skeptics. At the outset, the priority was placed on rolling out relatively simple improvements that could establish working relations with the most impacted villages and quickly demonstrate results. The company broke with the idea that its task was to fix problems caused by the dam by articulating an "integrated livelihoods approach" that also allowed it to focus on comparatively easy issues (such as water supply) or embark on big livelihood changes (such as reducing swidden farming). Water pumps were comparatively easy to build without extensive village involvement or any attempts to change villagers' habits; the results were tangible and provided a real benefit especially in villages that found it hard to access surface water due to the steep banks along the reservoir. Compensation for lost riverbank gardens was also possible. Caught up in the emphasis on speed was criticism of wasted money and time hiring foreign consultants, and the approach eschewed "analysis and paralysis," as the saying goes.[5] Was he supposed to use statistical analysis to identify bogus compensation claims? Did he need to relearn calculus to do this? He posed rhetorically. How would he explain to villagers why some did not get what they claimed? THPC simply gave families rice instead of sorting out direct cash compensation and got to work setting up irrigation pumps for new garden plots. "We had already been criticized for spending on consultants' fees and research—better to start now, make some mistakes and fix it along the way, than just do research." Speed served to demonstrate results quickly to dispel any notion that the company does not take its obligations seriously.

Delegate all or nearly all technical interventions to Lao management. John set up the program and served as its initial manager, but they soon hired Buali as the department manager in order to "eliminate the white man," as John claimed. The technical staff of eight full-time positions was all Lao, and the company took advantage of hiring master's graduates from the Asian Institute of Technology in Bangkok. Again and again the capabilities of the technical staff were insisted upon. They are not just "the Lao staff,"[6] Richard told me, deploying scare quotes, alluding to stereotypes of inefficiency and lack of initiative among development organization staff. Again the emphasis was on results and stimulating personal initiative among the technicians.

"We wanted to see 12–13 results go to market, so one person was assigned to maybe two results [i.e., activities] but mostly it was one-to-one. It's their responsibility to bring in short term contractors for design and set up [e.g., construction of water pumps]. [Our staff are] qualified people, [but without] a lot of experience." As activities shifted decidedly toward livelihood interventions, including for-profit production and the use of fertilizer, credit, and other forms of intensification, the demand for rapid results took form through Lao-specific sociocultural techniques for motivating or obliging villagers to get to work. On the one hand, the quality of these relationships was much harder to access by activists or the stray anthropologist who might want to understand how these forms of power played out. It was certainly not possible to claim these relations were an imposition of either bureaucratic state power or naïve foreign development expertise. On the other hand, the tasks were understood as complexly sociotechnical with explicit attention to the conditions of motivation and obligation through which villagers might be taught to be entrepreneurial.

Minimize expert research for specific biophysical issues. Systems were set up for monitoring erosion and water quality. Attempts by consultants to conduct more thorough investigation of flooding events, dramatic changes in the riverbed topology, or suspended solids were rebuffed, leading to consultants leaking unapproved research to activists. Fisheries received by far the most sustained attempt to assess conditions of natural variability and causality. However, even these were tortured affairs. The first expert report was scuttled when the consultant objected to attempts to change his conclusions. The second was long overdue, only to be met with criticism within the industry when it was finally released due to its narrow methodology and conservative conclusions, many of which argued the company would gain little by investing money in expensive possible solutions (minimum downstream release, aquaculture, a fish ladder to bypass the dam). While these conservative recommendations were probably accurate, the effect of the fisheries studies was to defer and delay any action for the fisheries problem, one of the most protracted and unresolvable issues for village livelihoods. In the meantime, the project experimented with a few aquaculture pilots (digging fish ponds stocked with expensive fingerlings fed a diet of pricey commercial feed) and livestock (pigs and poultry, with credit provided for start-up costs)—both protein substitution and market-based production to be trialed by a few select village entrepreneurs.

"Fisheries people think strangely," John told me. "They always want to figure out objectively what really happened, so they have to develop controls, figure out baseline estimates and so on. But the world is not a billiards table. We try to get livelihoods up to a level to have resources to feed, clothe and send children to school—how's that for an indicator!"

Channel socioeconomic monitoring into performance indicators. Since there had been no significant baseline research done before the dam was built, it was impossible to answer or even address the central question of the company's liability, namely, whether people were better or worse off at this point, six years after operations commenced. By 2005, there apparently existed scattered data collected in some villages, but this data remained unanalyzed and inaccessible. (The conversation I recount at the beginning of the introduction showed how hard it was even for industry consultants hired by the company to access the data.) At best the data went back three years. The effort was put into monitoring for variables such as malnutrition and school attendance that could serve as proxies, but in practice this information served as talking points for playing up successes that were impossible to substantiate—resulting in performative speech acts like the comment about trending the data with which I start the introduction. There was no comparative investment in rapid rural appraisal or other specific sociological techniques to design the different interventions or assess how they might play out in different villages. One could argue that such research was unnecessary, since the company had invested in an extensive plan characterizing the kinds of interventions needed, but there are many missing steps between an indicative timetable of activities and the technical investments needed to design and execute them. Those questions were rather dealt with informally and intuitively in dialogue among John, the technicians, and the villagers; through pilot programs; or simply through implementation followed with informal evaluation. Performance indicators effectively substituted for empirical research on causal relations.

Clearly separate transnational reputational concerns ("risk perception") from managing village-level interventions. English-language documentation was kept to an absolute minimum, primarily taking the form of a monthly report from Buali, the lead environmental manager John, and Richard; this report was only to include activities completed and any unexpected occurrences. This enabled Richard to control English-language interactions through his personable, verbal, and informal style, effectively keeping things off the record, while nearly all substantive interventions were organized by Lao staff in the Lao language and remained practically off limits to transnational activists (and difficult to access by foreign researchers such as myself). For instance, the performance evaluation became a paradoxical case in which the document report was completely unavailable for months, while Richard and John nonetheless cherry-picked some of its most positive conclusions to brag about in controlled, face-to-face settings.

Shuttle responsibility toward the market or broader societal trends; undermine the empirical basis of claims of liability. Many interventions were targeted toward making village economies self-reliant on market activity as a substitute for increasingly untenable core subsistence activities. Market interventions, including the entrepreneurialism that was taken to make villagers ready to absorb economic risk, formed a teleological let-out from the far more costly problem of reconstructing viable production, especially for the fisheries. Likewise, they blamed lower fish catches on population in-migration, which served to confuse questions of responsibility through a classic strategy of muddying the waters about who is to blame rather than insisting on the veracity of an alternative claim to truth. (After all, the in-migration was largely caused by the local growth due to the dam.) Ongoing sporadic flooding of riparian rice paddy, which threatened considerable areas of production land, was passed off as unrelated. Contract farming for tobacco and maize with private sector creditors was actively promoted.

The endgame of the management strategy was one in which market forces and population changes would muddle any disputes about corporate liability; this was combined with systematic disinvestment in research, undermining of potential claims of liability, and bulking up on feel-good talking points that performed commitment and goodwill. The Theun-Hinboun Power Company systematically worked to obscure its causal liability for socio-environmental impacts. It did so *both* through a mimetic performance of corporate responsibility *and* real work to ameliorate living conditions in affected villages. The problems that remained—long-term, unstable erosion; unpredictable flooding of production land; profound changes to riverbed topology; substantial collapse of fisheries ecology—were essential to riparian lives, and too big to fix without hampering the purpose of the dam. Performance-based management worked as a kind of subjugated knowledge of the business enterprise and the socioenvironmental relations that came within its orbit. The kinds of tactics that were put into place efficaciously undermined the ability of critics to challenge the company and served to de-emphasize empirical research in favor of energetic trialing of activities to see what worked.

Conclusion

As can already been seen, the environmental practices surrounding these anthropogenic rivers did very little to establish "facts on the ground." In her *Facts on the Ground: Archaeological Practice and Territorial Self-Fashioning in Israeli Society*, Nadia Abu el-Haj (2002) shows how the cultural practice of archaeology became the staging ground for territorial claims for the Israeli nation-state. Archaeologists shared an epistemic culture regarding how facts were constructed and ethnic identity assigned to artifacts in the material record. The a priori of modern Israeli sovereignty was made to rest on "texts, dates and pots" (104); the facticity of historical processes and cultural traits hinged on "low-level generalizations . . . built on the basis of things that could be seen" (129). In the political struggles of Lao hydropower, clearly one option might have been to make a hegemonic claim over the objective conditions of environmental damage and repair. However, a substantially divergent mode of practice, which required little empirical analysis and eschewed any explicit scientific objectification, was put to work instead. Taken within a genealogy of material practice, and following Michelle Murphy, anthropogenic rivers in Lao hydropower are "materialized as uncertain."

I remind the reader that this project is effectively a best-case scenario. The engineering and construction of the dam, with all of the accumulated facticity of expertise that it required, diverges sharply from routine disinvestment in science and practical expertise needed to manage late industrial environments. From the vantage of an anthropology of science and technology, undone science is best understood in light of the long-term disinvestment in any "knowledge infrastructure" involving repeated improvement and increasing accumulation of knowledge resources across divergent spaces and problems (Frickel et al. 2010). Investigating how sustainability sciences have pursued a long-term program of construction is beyond the scope of this book, and I can only point toward systematic research that has been done for agrarian development (e.g., Fernea 1969, Chambers 1974, Scudder 2006). But even in closely monitored situations, systematic expertise only weakly influences practice and implementation. No developer would leave construction of the dam to a program of strategic improvisation backed by good intentions, but this is common course for managing socioenvironmental impacts.

Meanwhile, it is tempting but incorrect to say that the dam itself *does* constitute a fact on the ground. The empirical reality of the dam with its permanence and contingent effects is certainly undeniable. As Richard was fond of pointing out, when criticized professionally for being too accommodating to activists, "we're built—we're not going anywhere." Does not the dam itself *exist*? Yet given the pervasive contestation, the ongoing, unstable transformations, and the delicate situations in which this project continues to take on new forms, perhaps the question is not *whether the dam is a fact* but *when is this dam?*. Sometimes, no doubt, it is a fact, but not always. The dam never just "is"—rather, its shifting temporality is an integral feature of *what* the dam is. Star and Ruhleder (1996) famously ask "when is infrastructure?" as a reminder that even the most solid reality has a history and is subject to decay, while every fact becomes deformed in time and risks appearing totally different from an unexpected vantage. If technologies temporalize, then the question of "when is infrastructure" is not only a reminder that the English-language verb "to be" carries with it an essentialist fiction of ontological fixity. More importantly, figuring out what temporality is instantiated by the technology in question is fundamental to understanding the technology itself.

Too frequently, sustainability concerns are chalked up as either effective or not and their representations as either misleading or not. Yet the task of anthropology in this case is not to judge effectiveness or the accuracy of representations. Rather, its task is to underscore the sociotechnical practices that hold these contradictions in abeyance. Anthropogenic rivers create novel possibilities for anthropos—and surely the strange creature of the manager must be counted among them, for it is the manager who contorts his body into an impossible shape to speak to such divergent interests. Looking at management in terms of its own logic shows that a degree of abandonment is one critical end point for privatized government (Povinelli 2011), and yet commercial sustainability management also invests substantial, real effort into addressing the problems it has caused. Perhaps it is a kind of "managed care," to evoke that mediocre form of health care in which "care" can only be understood in impoverished and ironic terms. Performance-based management as a concept is meant to hold open these multiple possibilities. Its rationality is therefore at stake in the ecologies it produces.

Furthermore, it is crucial to recognize that the risks borne by these sustainability managers are primarily the risks of failure and professional discredit, and only tangentially those of dangerous riparian ecologies and the multiple modes of living they enable. Clearly, while Richard and John articulated a long-term, open-ended engagement with these anthropogenic rivers, they themselves were living with rivers only within the terms of their professional engagement and never in direct risk of personal harm or the dissolution of a world in which living is possible. One cannot understand the willingness to fail or the arrogated right to muck things up outside of the historical privilege of white masculinities in which the real consequences of failure are borne by others. If I seem to condone or at least tolerate this experimental ethos it is only because I wish for vastly different experiments involving vastly different collectivities.

Richard's charismatic, personable engagement was not a performance of authoritative knowledge that insisted on rigid defensiveness. Central to his work was the open acknowledgment of problems and public engagement with what they were doing to fix them, which came as a welcome relief to critics. By the same token, his practice shared with activists many presuppositions about opportunism and possibility that had stymied previous attempts to manage those activists. Rather than succumbing to institution hacking, he turned the tables on activists and adopted a strategy similar to their own by trying to undermine their conditions for acting. In attempting to isolate the rationality of this mode of practice, I have identified an ethos of performative achievement to call out managers' familiarity, comfort, and even pleasure in working with uncertainty. Entrepreneurs thrive off the play of threat and opportunity. In turn, performance-based management produces yet more risk. It denaturalizes what was taken for granted and functions, at least sometimes, to actively disable conditions for life and work.

In particular, sustainability management, in this case at least, does not depend on the kind of authorized knowledge that is at the heart of classical understandings of knowledge/power. In chapter 2, I showed that the actor network concept had to be understood as a product of its times rather than an ahistorical concept that introduces a radical rejection of the modern distinction between nature and society. In this chapter I use Foucault against himself, not to show how apparently natural facts can be given a history but to show that reality itself becomes deformed in time, and that deformation can be an explicit strategy of subjugated knowledge. This approach has come out of an engagement with management practices on their own terms, in order to call attention to the fact that very often our assessments of neoliberalism have derived from the reason of economy and government, not the subjugated knowledge commercial logic of management. Deconstructing truth effects or the facticity of objects does not appear as the most crucial task in this situation. To affirm environmental justice here would be to wish for a lot more facticity, truth, and stabilized nature and to recognize the fundamental difficulties of composing those relations.

Sustainability performances or greenwashing are hardly new phenomena, but their apparently trite superficiality has protected them (see also Kirsch 2014; Ottinger 2008). Marina Welker (2009) deftly shows how corporate sustainability in Indonesia invokes quasi-environmental rituals, as when corporate leaders publicly swim in polluted waters to "prove" their safety or invest in marginally relevant participatory conservation programs to burnish their green credentials. Such ritual gimmicks are easy to criticize on rational grounds but harder to understand how they still work, regardless of their apparent superficiality. The analysis I present identifies some basic presumptions. If Richard's charismatic image management is any indication, these environmental tokens function as performative achievements that serve to disable, distract, and undermine rather than to establish the authority of a stable understanding of reality. They do not convince people of what is true, nor do they blind people by spectacle. Rather, they produce uncertainty, much in the way that climate science denialism is not effective because people come to believe climate change can be attributed to sun spot cycles, but rather because they come to doubt the veracity of mainstream science. These performative tokens, even when cheaply made, make it harder to ask important questions. They make it possible to change the subject. They give corporate managers a chance to be nice to neighborhood children. They give cover to industries to extract resources for as long as possible in a politics of deferral and procrastination. In this, they participate in what is fundamental to late industrial environments, for their destructive opportunism actively undermines the intelligibility of other worlds. Meanwhile, sustainability tokens are also part of our emergent anthropogenic ecology. They cannot be discounted, for they forge the ethnographic substance of day-to-day living and at some level they create and replicate dynamic forms of life.

If one looks at the kinds of anthropogenic ecologies that business actors are de facto in the position of managing, it becomes apparent that assessing business knowledge on its own terms, rather than as a kind of insufficient science, will be essential to understanding the actual ecologies that increasingly define our contemporary. There is a definite connection between Richard's performance-based management, the absence of reliable information about the effects of the dam, and the conditions of lived ecological uncertainty that increasingly characterize the environmental citizenship of Lao farmers and fishers living along these transformed rivers. Yet the empirical science of those conditions was never the motivation through which practices of environmental management proceeded. Richard and John's ecological practice targeted the assiduous, deliberate effort of activists. In doing so they constructed a risk management arrangement that both sought to ameliorate conditions for Lao villagers to a limited extent and made it exceedingly difficult to make straightforward claims about the kinds of harm they suffered. Neoliberal natures have long been understood to function not through the alienation of labor but through acts of dispossession (e.g., Swyngedouw 2007), and I have tested readers' patience by deferring, as long as possible, adequately rich empirical description of the dispossession of Lao villagers from their own perspective. Yet the reason for this should now be clear. Villagers' worlds have been remade by activist and managerial practice that they have no access to and scant understanding of its practical workings. If their worlds become unintelligible, it is through the articulated political ontology of these activist and managerial practices. Only by understanding the practical articulation through which this transnational assemblage has been composed may we hold open the possibilities—some of which are quite dangerous—of anthropogenic rivers.

It is tempting to say that rivers have the ultimate agency in this situation; that rivers' uncertainty establishes a pattern according to which social relations come to vibrate, regardless of the modes of reason through which people explain their ontological condition; or that the task should be understanding rivers' mode of existence without reducing it to anthropocentric frames. To a certain extent that is possible. But agency is not a zero sum game, and there is no necessary (determining) reason why activist or managerial practices come to vibrate with rivers' resonance, nor is it correct to say that the connection is simply contingent or accidental. Debates about human versus nonhuman agency must pay careful attention to the specific sociotechnical forms and commitments to rationality through which mutually constituted ontologies come into being. In particular, a claim for rivers' agency, in this case, could directly absolve the construction of the dam and the sustainability management procedures of their responsibility. One does not need to give developers and managers mastery over nature (or its rational comprehension) to understand their powers. They simply attempt to harness forces more powerful than they understand and bigger than they control. Methodological attention to how uncertainty is marked, grappled with, negligently ignored, and sometimes deliberately produced serves to remove agency from the knowing subject without absolving him or her of liability. It does so because it acknowledges that the subject is conditioned by riparian forces, not directly or immediately but only to the extent that commitment and obligation have been established.

Managers' carefully maintained ignorance, after all, is an essential component of their liability. Whereas the activists used their techniques of technical entrepreneurialism to produce uncertainty in the hydropower industry—uncertainty was an outcome of their practice—the sustainability managers went a step further to turn the deliberate production of uncertainty into a practice, tactic, and weapon in its own right. This in turn relied on a substantial history of management techniques, which, taken collectively, sharply undercut the dominate narratives of management in the social sciences. Management is the dominant practice of living with anthropogenic ecologies. It is the inadequate, hegemonic answer to Beck's (1999, 18) question—"How will we handle nature *after* it ends?" Development tools without development rules is not a theory that distinguishes effective private sector management from the clunky, bureaucratic and rule-bound efforts of NGO or government sector developmentalism—although it was clearly meant to function discursively in that way. More appropriately, it marks a domain of sovereign managerialism that claims for itself a kind of license and arrogates the freedom to work with impunity. As the narrative continues into chapters 4 and 5, it is important to understand that this relationship between activists and managers is decisive for the on-the-ground subjection of villagers. The sustainability enclave that took form in central Laos was an effect of the transnational assemblage through which Anglophone activists and managers came to have the greater say over conditions of living within these two adjacent watersheds.

As we will see, management happens after the event and therefore constitutes a biopolitics of partial abandonment or a pragmatics of letting life die (Povinelli 2011). The dam is already built. Its openness and its futurity compose a wake—in multiple senses ("awakening"; the wake of an event; the wake of an aquatic flow; the wake of a funeral; see Kingsnorth 2015). In this sense it is "after knowledge" (Strathern 1992) because, as a social form, it articulates a way to proceed after the collapse of trust in expertise and because some participants nonetheless still hope to find some mooring, some grounding, in the neutral representation of conditions of life along the rivers. This will be especially apparent in chapter 4, where I look in more detail at the performance evaluation Richard tried to establish with IRN. Finally, if the management of stasis and neglect is one element—not the only element—of the telos of sustainability management, then this raises the biopolitical prospect of letting life die. These are the concerns of chapter 5, in which I think through the chronic problems of living with anthropogenic rivers.

Interlude

The Method of Uncertainty

Rather than an ethnographic vignette, in this interlude it may be useful to delineate the variety of ways in which I use the term "uncertainty" at different points in my argument. For it is clear that I was *taught to see* uncertainty by analyzing the practices of environmental managers while asking Foucauldian questions about what characterized their approach to knowledge. As previously emphasized, the essential ethnographic problem was that they routinely rejected a detailed empirical basis for formulating plans of action and, in contrast to authoritative speech acts more familiar in expert systems, these industry actors were also comparatively more willing to say that they did not fully know how to characterize socioecological issues. I then deployed uncertainty as a methodological entry point to investigate practices surrounding anthropogenic rivers. Lastly, I make specific claims about the production of uncertainty as a practice, its status as a component of knowledge (rather than just lack or limit to knowledge), and a constitutive element of late industrial environments. Since the concept of uncertainty derives from an ethnographic analysis while my use of it significantly exceeds any colloquial or "native" understanding, my approach follows similar work that treats vernacular concepts or ontologies as deserving of sustained attention while pushing their conceptual possibilities to the limit.

The idea that late industrial environments are constituted by uncertainty is an ontological claim. It is certainly possible to understand late industrial environments as impoverished or damaged from a perspective of ecological function, biodiversity, or socioeconomic utility. So too it is possible to criticize ecological destruction on aesthetic or moral grounds. However, I find the idea of ecological change as a simple *reduction* in complexity or value to be insufficient. By claiming that anthropogenic or late industrial environments are constituted by uncertainty, I underline a number of implicit corollaries.

First, ecological destruction by itself is insufficient to account for the profound productivity of contemporary ecologies, even if that productivity appears bleak or negatively valued compared with intact ecosystems. In historical timescales, destruction does little to help us understand vastly different disease vector populations that characterize industrial farming (Fearnley 2005; Keck 2008), the ways certain "pest" or "invasive" species thrive in damaged ecosystems (Raffles 2017), or the ways depopulated disaster landscapes become habitat for population booms (Davies and Polese 2015; IAEA 2001; Tsing 2015). On geological timescales, we can expect that the high rates of biodiversity loss currently underway will result in major evolutionary transformations, just as we must acknowledge that climate change will result not just in an apocalyptic wasteland but an Earth that from our present understanding will be fundamentally unrecognizable. There is no need to unguardedly celebrate these anthropogenic ecologies in order to emphasize the dynamic and radically open futures they imply; it is certainly clear that existing institutions will not protect us from those dangerous futures. Even if we work now to make those institutions stronger and better, my position suggests we treat that as an experimental and creative activity rather than an attempt to adhere to predetermined political and scientific forms.

Likewise, anthropologists can and should reject the political, ethical, and aesthetic consequences of these changes based on a vibrant ecological politics. By claiming that unpredictability is an integral part of the ecological changes underway, we are better able to reject certain changes and resist their anthropogenic causes without succumbing to the desire to know for certain what those futures may hold or demand ecological purity within a predetermined vision of what we currently feel is the correct ethical principle or scientific understanding of ecology. For example, too often those who protest hydropower dams are forced into the position of claiming they know what harms will result from the dams in question. But that is unreasonable. Those who develop the potentially dangerous infrastructures of late industrialism should always be put in the position of producing reliable, quality research about the potential effects of those projects; the knowledge used to assess and mitigate environmental impact is far too inadequate, with scant infrastructural underpinnings, meager oversight, and totally insufficient funding. Harm or environmental damage is only part of the problem, for uncertainty itself is a major negative effect forced onto people affected by these projects. By articulating uncertainty with all of the deeply resonant fears, dilemmas, and hopes that it holds, environmental protest is better able to escape the strict economism of a cost-benefit mentality that accompanies the language of damage. (There is also much work to be done to understand how ecological value does not uniquely derive from human assessments, and whether all living beings directly evaluate the relations in which they live.) Developers do everything they can to limit risk to their financial returns; we need a thorough assessment of how that risk gets pushed onto other people and living ecologies as a direct impact of the development process itself.

Anthropogenic ecologies are shot through with embodied, physically instantiated knowledge that contains within it long genealogies of thought and practice, much of which is fraught with severe limitations. It is important to recognize that environmental and sustainability knowledges are vast in their scope and essential for many ways in which human-environment relations are mediated. These environmental knowledges take all kinds of forms, from waste water treatment to wildlife population management to resettlement procedures for large dams. By emphasizing uncertainty as a critical component of this relation, observers are better able to show how these knowledge infrastructures rarely have the degree of integrity and capital investment enjoyed by many forms of higher priority knowledge. These forms of knowledge are socially devalued and underinvested. By the same token, too often we tend to see engineering or technological knowledge as separate from environmental knowledge because environmental knowledge is narrowly framed as pertaining to organic things such as animals, forests, or landscapes or human relationships that are close to these things. But in order to understand the vast ecological changes underway it is necessary to understand apparently nonenvironmental knowledges as both primary drivers of ecological change and as containing narrowly construed or limited ecological components in their own right. To characterize an industrial plant as fraught with ecological uncertainty makes clear that the industrial process is always thoroughly integrated into an ecology and that the knowledge that underpins it is wholly ecological regardless of whether the engineers see it that way. Engineering is always already ecological, even if the ecological knowledge at work approaches nil. More provocatively, what would it look like to design and build dams with a highly integrated, rich attention to creative ecological design and collective decision making? What if the ecology of large dams was not viewed as a limit or prohibition, but as the centerpiece of a collective process that both affirmed flux without mastery and was backed by real investment in ecological knowledge? Inadvertently, the tendency to see environmental change only in terms of destruction results in a critical politics that, at worst, can only reject projects that get built anyway and, at best, can suggest modest technical improvements to a design process that considers ecology as a secondary concern.

At a degree of remove, these ontological claims give way to a historically contingent "phenomenology" that helps draw out lived experiences of late industrialism by showing that temporality is integral to infrastructures themselves, including knowledge infrastructures. In debates about environmental damage resulting from large projects, the usual narrow focus on specific, identifiable harms does not do enough to characterize people's experience, and results from a misplaced literalism and demand for objectification of discrete quantifiable harms. For example, debates around this dam often dealt with whether the dam had caused increased flooding or rice paddy loss, making that paddy too risky to farm. But even if the cause-effect relation of the flooding is well understood, the inability to farm that land is only the first layer of consequence for farmers and their families. When debating whether to plant a risky crop, they are also weighing whether they should instead send their son or daughter into Thailand as a seasonal migrant worker (and what kinds of uncertainty it entails in turn); they are wondering whether they can make needed investments to their home or farming equipment if they are witnessing a permanent decline in arability; they are debating whether to get involved with contract farming, which represents a completely new debtor relation. In short, the flooding calls into question their future—not as an established fact but as a murky domain of the unknown. There are many examples of similar temporalization effects and an anthropology concerned with the historicity of experience can do much more to understand that ecological relations are not immediately present but rather constantly deferred, anticipated, and imagined, and that ecological imagination incorporates or puts to work distinct forms of knowledge ranging from vernacular know-how to expensive forms of scientific knowledge to all sorts of nonpractical knowledge. By emphasizing the uncertainties of these environments it is much easier to render the ethnographic experience of late industrialism through a description of its phenomenal experience.

In practice, capitalist actors routinely produce uncertainty not as an unfortunate by-product but as a deliberate tactic of disruption that makes it possible to disorient, undermine, confuse, or defer political challenges to their industry. Whereas late industrial environments are constituted by a productive uncertainty (independent of any agent's deliberate practice), the production of uncertainty in this sense refers to a deliberate tactical orientation. Efforts to undermine environmental sciences and regulatory research fall into this category, as does disingenuous greenwashing and sustainability eco-hype. This domain of practice must be situated within a genealogy of techniques that form a tactical repertoire; it presumes deliberate, motivated activity by human practitioners whose agency cannot be distributed wholly to nonhuman actants; and this activity, finally, should not presume any mastery or even the subjective belief in mastery over people and things. Here the experience of uncertainty opens onto a political ontology in which, implicitly at least, actors understand uncertainty to be part of the warp and weft of political relations in which their adversaries are able to operate. In other words, creating chaos is not just an accident or by-product, but part of the calculation of a strategic relation to what exists. If an actor relies on a tactic of disorienting her adversaries, then there is an implicit ontology that acknowledges and affirms uncertainty as essential to distributed political capacities. Moreover, since these actors routinely act without extensive research or empirical assessment, they betray an affirmation of uncertainty that is clearly distinct from the classic formulation in which risk stymies authoritative decisions.

Furthermore, by tracing an experience of uncertainty and the way it posits a political ontology, anthropology will be better able to understand how things are done to people that they have no basis for understanding, for what is being done in circles of finance, planning, or government is frequently outside the scope of their experience. Uncertainty in such a case is closer to a kind of systemic ignorance in which thoroughly opaque, complex, and inaccessible processes have decisive effects over the lives of people excluded from their operations. In fact, people do often have some partial, speculative understanding of what is going on (or an awareness that something is going on even if they do not know what exactly). In such cases, the challenge is not to describe an experience or language of uncertainty, but rather to tease out the partial connections between hierarchical domains of practice and demonstrate the connection to what people think and feel on the one hand and the structuring conditions of systemic ignorance on the other. For example, people affected by hydropower development must figure out how to navigate the complex bureaucracies and expert knowledge systems that manage and discipline their lives while frequently excluding them from access to compensation. The uncertainty or ignorance itself is an essential part of the experience and the act of disempowerment. Like the systematic underinvestment in ecological knowledge characteristic of late industrialism, in this case uncertainty is not only part of the give-and-take of political practice but comes to form a hierarchical structural condition.

From an epistemological standpoint, uncertainty is a useful method for the ethnography of expertise because paying attention to how experts talk about what they do not know leads the ethnographer directly into the core epistemic dilemmas of their practice *and* the sociopolitical motivators that push their knowledge practice in certain ways. Uncertainty is the name given to limits of expert knowledge; the commitments, care and curiosity directed at objects of expert knowledge; and the social demands placed on experts. Focus on the uncertainties of expertise puts into direct dialogue the social relations of expertise (including relations with things) and their core epistemic dilemmas.

The approach taken here—especially in chapter 4—is to stay close to the dilemmas posed by expert consultants themselves as they engage in sociopolitically meaningful research. This requires accounting for ideas, practices, and dispositions as they come up within a body of historical practices without feeling the need to provide a comprehensive genealogy of, in this case, environmental mitigation consulting. Events place demands on experts, and uncertainty helps track how those events push their knowledge into uncharted territory where it is forced to improvise. It also lets me stay close to the issues that matter to the experts at hand, to show how they deliberate on these issues within a body of historically specific practices of reasoning, to track their improvised and imaginative ways of engaging with novel environments without assuming a totality of systemic knowledge, and finally to show how knowledge practices interlace with a variety of social forms, from ethical reflection to the contours of professional consulting to gendered power relations. As with the experiences of those affected by major projects, attention to experts' uncertainties keeps the ethnography close to their experience of the ecological event while documenting what kinds of demands the event provokes in the broader network of relations in which they are enmeshed. Uncertainty tracks the remainder or excess of established knowledge and the ways it becomes subject to critical rethinking; studying the boundary between what is known and what is unknown assesses the reality of the event without assuming consensus or a fixed paradigm.

The epistemological concern with experts' uncertainty bears on my more fundamental claim that uncertainty is a relation that indicates the value of knowledge. That is, I understand knowledge as a kind of relation rather than a thing that one possesses. Inevitably fraught with uncertainty as a constitutive element, knowledge binds the knowing subject within a complex material-semiotic web rich with histories, curiosities, skills, practices, and speculative imagination. This is a claim about how we might assess uncertainty without talking about it as a nebulous, apparent nonentity, a lack, absence, or limit of that "thing" called knowledge (as indicated by its etymology). If knowledge is not a "thing" then uncertainty is not its absence. Rather, uncertainty is an ever-present constitutive element of knowledge that indexes the value of that knowledge to projects of life and living. In fact, the simple ability to articulate what one needs to know—to ask the right questions—is a powerful act.

Chapter 4

The Ethics of Document Engineering

Charles was an environmental consultant with long-term experience in Laos, who was American, retired, and living in Bangkok. We met up in his modest townhome in a comfortable neighborhood away from the glossy lights of Bangkok's city center. He had worked in the hydropower industry as an environmental consultant during the 1990s, when the industry was trying both to get dams started as rapidly as possible and to fend off critics and activists who increasingly found ways to challenge experts such as himself, who previously could speak authoritatively for the environmental and social problems that might be caused by dams. "Writing these reports is an art," he told me. "The environment is never a problem unless someone thinks it's a problem. Environmental problems are always social problems." Charles challenged my assumptions about the depoliticization of expertise and forced me to think differently about the task of consulting from the vantage of writing itself.

Charles described to me how he hurriedly completed a draft environmental assessment report for a highly controversial project so that he could leave to manage some impending family matters in another country. His replacement, a junior staffer working for the same consulting firm, was called on by the project manager to edit Charles's draft document because many of its conclusions seemed ambiguous. The project manager wanted the document to clearly state that social and environmental impacts would be minimal. He wanted the environmental report to offer a strong recommendation to commence with financing and construction. Sitting down with the junior consultant, the manager went through the draft document crossing out ambiguous phrases and replacing them with clear, favorable recommendations. The manager wanted to ensure that there was no ambiguity at all: the environmental report should boldly claim that there would be no controversial environmental problems for activists to latch on to.

At the time, Charles had taken his own copy and mimicked the activity of the manager, crossing out his own words and writing in the words of the project manager—crossing out his own ambivalent recommendations and filling them in with decisive prose. It was this act of crossing out and physically replacing the words that held the consultant's attention as he retold the event.

Yet the catch was, Charles told me, that the manager's decisive recommendations inserted into the final draft were bound to raise eyebrows. The project manager was insensitive to the ways documents like this were increasingly taken up by environmentalists and others and read precisely as biased and political rather than disinterested expert reports. As the draft went on to the World Bank for comments, Charles was vindicated. As he described it, "You can't just write things like that, it's much better to be a little ambivalent, *we find no reason for the project not to proceed*—things like this. Otherwise, it's like a flag that you're hiding something. In the end he went back to the version I submitted."

With respect to his own work as an engineer, Charles described his labor not as engineering the physical structures he was trained to design but rather as engineering reports that did certain kinds of work within a professional, institutional context. Environmental assessment work was not a science, he said, but rather an art of managing relationships within the text. He even claimed that creating environmental reports was "great fun," involving a challenge of "pitching" a narrative that "could be a novel that thousands read, instead of something used only by those what have to build on it in the next stage." Charles suggested that an environmental report could be read for its compositional value if only one knew the context, which the authors themselves must anticipate, project, and conjure into presence. Put differently, for its experts, in the writing of a document it is often possible to hear the "pitch" of environment—its tenor and its wager.

Ethics of the Anthropogenic

What commitments and expectations permeate the prose of expertise? What, moreover, are the obligations of expertise when written with an eye toward multiple layers of interpretive reading? If a development document must be shrewd—if it must calculate the political valence of its possible legacies—then the document can be taken up as an ethnographic artifact, as a commentary on extant relationships, a forecast of risky possibilities, and an attempt to manipulate or manage its anticipated milieu. Insofar as expertise is a labor relation meant to handle certain kinds of unresolved problems, an ethnography of expertise can reveal how political questions become taken up within an ethics borne privately and professionally by the experts in question.[1] Indeed, as I show in this chapter, the commitment and compromise of document engineering configures the very boundary of professionalism. In this case, at least, environmental expertise can be understood as a mode of affirmation, knowledge, and critique that answers to environmentally critical conditions even as its answers are inadequate.

The practice of collective writing of a performance evaluation was one primary way that anthropogenic experiments took ethical form for experts whose professional task was to produce socioecological knowledge. In this chapter I further explore the problematic of uncertainty in situ, the experimental form through which the company and activists briefly explored new ways of addressing its environmental obligations, and the problem of figuring out questions of value and evaluation in an unstable professional context. The central concern is, to what extent do anthropogenic ecologies provoke an ethics?

In Michel Foucault's research on ethics, the methodological emphasis on the concept of problematization enabled him to move away from earlier work in which the discursive episteme played an organizing role in his theories (see especially Rabinow 2003; Rabinow and Rose 2003). In particular, emphasis on ethical problematizations allowed him to specify the contingent connections between how a problematic of living is defined, which is a historical matter of designating what is important within an ethical formation (what he terms a mode of subjectivation), and the "ethical substance" through which it is possible to engage in ethical work (Foucault 2005, 1997a, 1997b). Desire, for instance, is the key ethical substance for understanding transformations of sexuality in Christianity and later as a problem of the modern subject, because it was desire that made it possible to do ethical work on the self; in turn, the "reality" of desire, or its truth if one prefers, emerges precisely in relation to that ethical formation. This reality is a philosophical problem to the extent that ethics bears on an "historical ontology of ourselves" (Foucault 1997c).

Hence, there is an implied double ontology at stake in which the historical ontology of the subject emerges in relation to the contingent historicity of a nonsubjective reality. "Problematization," Foucault writes, "does not mean the representation of a pre-existent object nor the creation through discourse of an object that did not exist. It is the ensemble of discursive and nondiscursive practices that make something enter into the play of true and false and constitute it as an object of thought (whether in the form of moral reflection, scientific knowledge, political analysis, etc.)" (quoted in Rabinow and Rose 2003, see also Foucault 1990). Distinct from any ecological concern, one could see this as an inherently anthropogenic movement: the historical ontology of ourselves must be situated within an historical ontology of an ensemble of material-semiotic relations.[2]

In this chapter, I show that professionalism is the *substance ethique* of environmental expertise. As a formalized and durable mode of practice, it allowed experts to confront the biopolitical stakes of late industrial environments, and it formed the terms of improvisation in situated relation to the dam's ecology. Furthermore, the risky commitment to an experimental form in this case underlined the novel environmental obligations the company had made, without presuming those obligations could be taken at face value. While standards of professionalism worked to convert political questions into matters of personal ethics, I show that the ethical is not a "private" concern segregated from professional practice, but rather professionalism draws directly on the ethical, and in ethical terms it construes new obligations to anthropogenic ecologies. I argue that one figure of ethics for anthropogenic ecologies is a concern to take seriously others' uncertainties. *Whose* uncertainties are acknowledged becomes the central question. Another, essential to these consultants' negotiated practice, is a concern that expertise be a form of criticism that is capable of being heard. The emphasis on technical criticism in the anthropology of science[3] should be tempered by a more subtle examination of the ethics of critique within technocentric and power-laden contexts. Not all technical criticism can be construed equally.

Experimental Form

Clearly, if the success of activists' initial advocacy campaign was to be at all meaningful, they could not take Richard's word that his company would fix the problems it had caused.[4] Indeed, no one could, and the question repeatedly asked of my research—"is this dam any better than the others?"—was on the lips of many expatriate development workers in Vientiane. More than simply being a matter of disputed fact, the *question* sits at the heart of what it means to practice sustainability, for it implies constant negotiation over the meaning of environment and the interplay between competing representations. This chapter looks at the company's experimental collaboration with IRN, which set the terms for a joint evaluation of the company's environmental performance, under the assumption that only an explicit collaboration could produce a stable perspective for evaluating empirical claims. Underscoring the open-endedness of the situation, the experimental collaboration was one of the main ways that the anthropogenic river came to induce new forms of being human, precisely by calling into question the stabilized boundaries between industry and activist actors.

In chapter 3, I described how cooperation with activists served to fold them into the company's very deliberate image-repertoire, so that it could be seen as making an effort and reaching out in good faith, essentially changing the game for activists by changing the negative tone and affect surrounding the conflict. But the activists had their own reasons to try out an experimental way of working with companies in a less combative mode. Strapped for resources, it was costly and time-consuming to conduct field evaluations to follow up on a single project while coordinating multiple campaigns around the world. A "tit-for-tat" dynamic had emerged around empirical claims about social and environmental effects. Finally, it was practically inconceivable to do the kind of detailed study of project effects when the hydropower firm and the Lao government would have to consent to consultants' presence in the field for extended periods of time. It is one thing for an activist researcher to spend a week or two traveling among villages asking questions while on a tourist visa; it is something else altogether for a team of researchers to spend a month engaging in more detailed interviews and analysis of a broad range of specific issues, including examination of sustainability activities and programs. IRN also took risks by becoming involved with one of the companies that caused such ecological damage. Environmentalist allies cautioned them against participating, and IRN also feared being coopted or alienating key people in their network; the stakes for experimentation on new ways of evaluating environmental knowledge were high for all involved.

The question was whether this collaborative experiment could produce an independent assessment of the sustainability measures. Even with the activists' input, the hydropower company would still be hiring the consultants. The conditions of the collaborative agreement were threefold:

(1) IRN would provide a list of acceptable consultants to be hired for the performance evaluation, from which the THPC would hire one each for agricultural, fisheries and social specializations.
(2) THPC would vet the terms of reference with IRN and reach an informal agreement about the basis of the evaluation, with THPC retaining control over it.
(3) IRN would have the opportunity to comment on the consultants' rough draft of the report, and consultants would address those comments in the report as appropriate.

On the basis of those terms, THPC hired three consultants to come evaluate their environmental work—a Thai social specialist in her thirties, Umporn, with experience in gender analysis and who had trained in Australia with a noted environmental research center; a British fisheries specialist, Steven, also in his thirties, with extensive experience across the Mekong with participatory activist research in Thailand's Pak Mun Dam and with good Lao- and Thai-language skills; and Douglas, a retired Canadian rural development specialist originally trained as a geomorphologist but with decades of development experience in Indonesia and Nepal. Over the remainder of this chapter, I show how deliberation about the form of evaluative practice, through their work of collectively writing a controversial environmental report, demonstrates the obligations and constraints that come to bear on the production of environmental knowledge. As previously described, their work was bracketed by a powerful schism that emerged when a rumor spread around town and Richard was forced to abandon the agreement with IRN, throwing into relief the tenuous aspiration of what they were trying to do.

Umporn and Steven had taken this job because of their interest in the collaborative process with the activist group, and in particular Steven was professionally and personally very involved with fisheries and rural poverty activism in the Mekong Basin. His activist loyalties lay with villagers and what he called "living aquatic resources," not just for these rivers but also for other rivers that might be damaged by future dams. The objective of forming a collaborative agreement with the activist group was to give the consultants and their report a degree of independence. When the schism emerged among Umporn, Steve, and the company, the first two conditions of the agreement had already been completed, but the last condition was never fulfilled, abrogating the overall objectives of the collaboration and drawing criticism.

The attempt at formulating such an experimental collectivity produced results striking in their own right, and the joint performance evaluation had taken on a life of its own. As the consulting team threatened to fall apart, Douglas, retired and the most experienced of the three, wondered what Richard had gotten himself into by getting involved with the activists. "I thought it would have gotten much worse. My big surprise was that Richard got himself into this situation. He's really smart, and I don't understand it. To bring three people down here and try to control them—I can't figure out why he would do that. He doesn't owe IRN anything."

From the perspective of the dam's managers, these consultants' final report would provide a clear vindication of their sustainability strategy with detailed advice on improving their operations. From the perspective of activists, the report failed to address the dam's most outstanding problems and largely ignored the fundamental conclusion that the project had undermined the villagers' aquatic and agrarian livelihood basis. But if neither the company nor the activists achieved their strategic goals, the consultancy still produced considerable constructive knowledge about the dam's impacts and operations. Bolstered by the professional ethical commitments of the different consultants, the collaborative and sometimes conflictual basis of the project review team made for wide acknowledgment of the dam's problems and the limitations of the company's approach. A reader who is willing to acknowledge the risks and practical entailments of what at first glance look to be mere practical recommendations will readily understand that the project faces severe hurdles to achieve an equitable outcome for the villagers.

However, my concern here is to understand expert evaluation as a kind of ethics of the anthropogenic. I use the process of negotiating and writing this performance evaluation document as a way of exploring the ethical relationships at stake as an outcome of obligations to the anthropogenic river. Documents have stories, and exploring the relationship of the artifact to the social process of its production and manipulation is one way to investigate the work of knowledge and its inherent uncertainties.[5] The substance of these ethical debates as well as the form of the collaboration are some of the main ways that experts came to experiment with the anthropogenic dynamism of the river.

Anthropogenic Ecologies

The difficulty of knowing anthropogenic ecologies saturates environmental knowledge and points in two complementary directions. On the one hand, the conditions of inadequate knowledge in any given ecological predicament demonstrate the constitutive underdetermination that arises when technological interventions produce effects that are neither determined from the outset nor accounted for within a preexisting body of science. On the other hand, the relation to inadequate knowledge is a critical feature of the kinds of self-forming ethical practices through which anthropogenic ecologies produce different ways of being human. After all, there are different ways to relate to the contingency of one's knowledge. Far from retreating into language or discourse as the only accessible reality, only a reflexive analysis that attends to epistemic limits (as any ethics of expertise must) is capable of a ontology open to its own contingency. By the same token, assessments of late industrial ecologies necessarily rely on the limited and partial knowledge of scientists and experts even if that knowledge must constantly be provincialized. Anthropogenic ecologies and the uncertainties that plague them dominated the concerns of consultants' work and formed a primary "mode of subjectivation" through which the subject of environmental expertise was construed in relation to the historical realities of the dam and its aftermath. Here I situate and discuss some of the conclusions of the report through an analysis of the company's work editing of the rough draft; it brings us ever closer to the ecology of the transformed rivers.[6]

Erosion impact on fisheries. In the report, there is a strong emphasis on monitoring erosion caused by the large amounts of water released from the power house. The loss of fisheries is perhaps the largest socioecological impact of the dam, and while the firm has been gathering data since beginning operations, virtually no mitigation work has been attempted. Because of the project design, a large amount of water (110 cubic meters per second at peak operation) is introduced from the Nam Theun River into the much smaller Nam Hai/Hinboun. By way of comparison, early project documents estimated that, during the dry season, the Nam Hai normally carried about seven cubic meters per second. The persistence and repeated fluctuations of water level have caused precipitous bank failure along the Nam Hai. The report estimates that 270,000 tons of sediment are introduced in the river each year. Umporn, Douglas, and Steven felt that the company, overall, had been wise to not attempt any mitigation efforts because those previously proposed would, in retrospect, have probably failed. The primary ecological impact of the erosion discussed in the report is the effect of suspended solids on aquatic life due to the reduced sunlight penetration and the filling of deep pools with sediment. By reducing photosynthesis, fish populations suffered.

The erosion was not going away. Two years after this performance assessment, another team of consultants worked to get the hydropower company to acknowledge the impact of erosion on downstream flooding and the collapse of wet-rice agriculture. The company rejected their diagnosis and the consultants withdrew from the contract. The problem had shifted, however. The river had become much shallower as it neared the Mekong, and, during peak flooding, the rice fields along those low-lying plains were inundated for far longer periods of time. Villagers I spoke with had ceased to plant many of their fields because it was too risky. According to farmers, the problem was the increased duration of flooding as well as the sediment carried in the water, which greatly exacerbated rice plants' susceptibility to infection. If erosion had seriously constrained fisheries production, it too appeared to be affecting the wet season rice that had long formed, with fish, the basis of the area's economy. No wonder many were progressively marginalized by the village credit and rice intensification scheme.

Intensified dry season rice cropping. The draft report correctly identifies the company's risks of pushing for high-input, high-yield dry season irrigated rice along the lower river. Promptly after the consultants finished their work and left, that season's crops failed dramatically. Villagers accused the hydropower firm of giving them bad advice and low-grade fertilizer, and it did not help that most were now seriously in debt for the fertilizer, pesticides, and seed they had purchased using newly set up village microcredit schemes. The backstory is that these villages, like so many villages in Laos, rely heavily on wet season, rain-fed rice, without fertilizer and with high labor inputs. These intensification efforts were simply unfamiliar, financially risky, and hard to come to terms with. But to make matters worse, the wet season rice for many of the same villages was to fail some six months later—due to flooding exacerbated by the erosion and attempts to stabilize shifting agriculture.

Stabilization of shifting agriculture. The report recommends continuing the company's attempt to stabilize shifting agriculture around the headpond. The hydropower company had encouraged the district government to implement land use permits in the area, restricting cultivated land access to three hectares per household—far too little for the necessary annual shifting crop rotation. The soils in this area, a kind of sandy moraine, were very poor. The firm had been paying farmers (in rice) to convert their recently titled swidden plots to poldered rice fields, meaning that they would dig the hilly land into "terraces" meant to capture water and enable the more productive wet paddy rice instead of dry hill rice cropping. Some months after the review team left the country, I was in these villages as their crops were failing miserably. In one vague passage in the report, there is an allusion to a village rebellion in which a large number of households from one village rejected the rice intensification project and instead turned to illegally pioneer slash and burn cropping in a protected forest. Two more villages organized more or less explicit resistance to THPC's environmental program later that year, inviting strong-arm intervention by the district government. They were obliged to participate by household—except for a few wealthy villagers who somehow got out of it. Not surprisingly, too, the impetus for this development activity was as much about protecting the firm's investment as it was about improving villagers' livelihood. Agriculture along the headpond was held to contribute to erosion and siltation of the small reservoir, a process that could undermine the long-term profitability of the dam.

Marginalized populations. Deleted from the final text of the consultants' report is the following: "Some resource poor farmers were unable to participate in the program this year because they did not repay their debt to the Savings and Credit Fund even though their rice crop was excellent. Quite often it was these same marginalized villagers who were most affected by the fisheries decline." In an edited version with the managers' annotations is a comment by the hydropower firm: "The first sentence is incorrect. Some farmers decided not to participate in the program. [The sustainability program] has never prohibited farmers from this program. The second sentence is not related to this issue, and thus, inappropriate." The report's authors here allude to a series of compound issues relating to indebtedness and marginalization. The deleted text says that villagers were "unable" to participate, but the managers' comment interprets this as claiming they were "prohibited" from participating. Clearly, the terms of participation in a loan program are complex and, especially in rural development work, can be usurious. Here and elsewhere the consultants emphasized that poorer segments of the villages often are on the sidelines, and they describe the compound nature of marginalization and impoverishment. Instead of trying to contend with the difficult problem, the managers cry inappropriateness. In the subtext, marginalized farmers sometimes manage to farm, but only just barely, even while they could no longer depend for their livelihoods on fisheries now destroyed by the dam.

Overall decline in fisheries. What fisheries interventions? The project had mostly not attempted to fix the problems caused for the fisheries. Like the issue with erosion, their approach was largely to wait and see, and the emphasis on "studying" or "monitoring" these problems raised considerable issues with the consultants. Douglas had commended the firm's "monitoring" approach to erosion; whereas Steven was troubled by the approach to fisheries monitoring—it was haphazard, they had rejected one detailed study because they disagreed with its conclusions, and they withheld another long-term study on the grounds that the revisions were not finished. In his contribution to the performance evaluation, many of Steven's comments were heavily edited by the firm because they imply the company had not done nearly enough to deal with the fisheries issues. At first glance it looks like a classic case of depoliticization in which problems that cannot be addressed are simply eliminated from the bureaucratic process (Ferguson 1994). Jonathan Crush writes, "In a textual field so laden with evasion, misrepresentation, dissimulation and just plain humbug, [development] language often seems to be profoundly misleading or, at best, of limited referential value" (1995, 5). And yet this report is a strikingly rich document. This technology of manipulating the physical artifact of the document is not the bracketing of politics described by Riles but the function of Microsoft Word that tracks changes in the editing process (figure 4.1).

Yet I want to argue something more subtle than depoliticization is at work, which has to do with the intractability of anthropogenic environments. While the dam's impacts on these villagers and rivers have been summarized in previous chapters, only here does the clinical, interventionist quality of environmental practice begin to come into view, drawing parallels with medical knowledge, which is both normalizing and normative (the latter in the sense of establishing new norms under dynamic conditions of living, see Canguilhem 1989). These stories provide only a first blush of the diagnostic, power-laden, and pragmatic dimensions of sustainability practice. Especially within this experimental domain, environmental knowledge established criteria that were demanding and powerful just as they were vulnerable to criticism.

One can sense hope and anger in Steven's prose, delimited in those blue boxes off in the margin, literally and graphically rendered hard or impossible to read (figure 4.1). That affect points toward the political economy of Lao hydropower development, identifying its destructive mode and the limits of fixing the problems that proliferate in its wake. The consultants' ethics do not take place in the absence of power relations, but rather the ethical is one mode of practice through which power relations are actively engaged and contended with. The ethical problematic, furthermore, is not a "private" concern in strict opposition to the professional conditions of expert labor, but rather it draws directly on the obligations toward anthropogenic rivers taken up and grappled with by environmental professionals. I turn to the evaluation of environmental obligations as a central domain of the ethical, for the consultants' work proceeded in what was a tricky situation. The consultants had come to realize that only the activist NGO was taking the initiative to monitor the firm's environmental compliance, performing a de facto regulatory role from its offices in California, whereas only the hydropower company was actually engaged with the villagers in trying to improve their

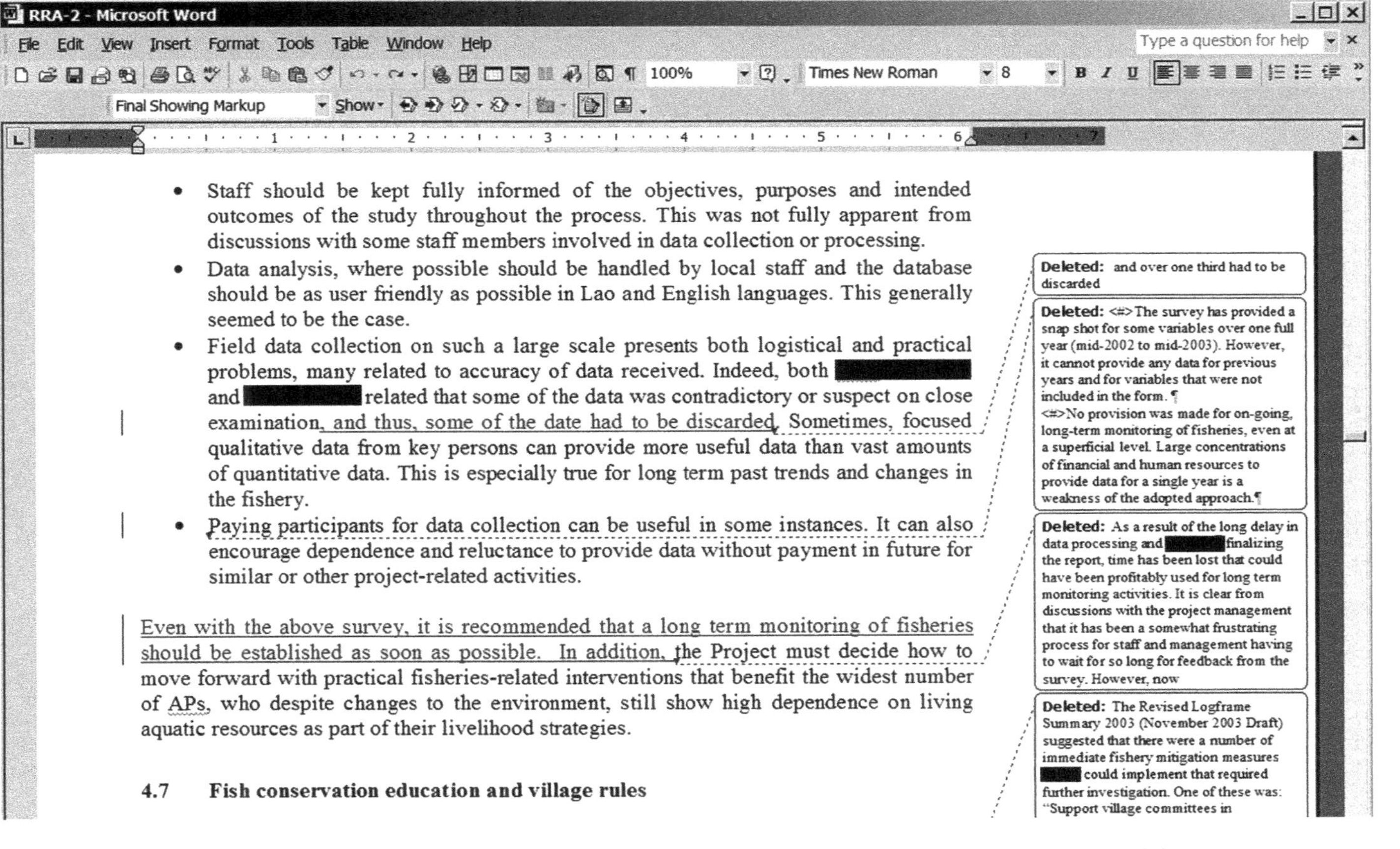

RRA-2 - Microsoft Word

- Staff should be kept fully informed of the objectives, purposes and intended outcomes of the study throughout the process. This was not fully apparent from discussions with some staff members involved in data collection or processing.
- Data analysis, where possible should be handled by local staff and the database should be as user friendly as possible in Lao and English languages. This generally seemed to be the case.
- Field data collection on such a large scale presents both logistical and practical problems, many related to accuracy of data received. Indeed, both [redacted] and [redacted] related that some of the data was contradictory or suspect on close examination, and thus, some of the date had to be discarded. Sometimes, focused qualitative data from key persons can provide more useful data than vast amounts of quantitative data. This is especially true for long term past trends and changes in the fishery.
- Paying participants for data collection can be useful in some instances. It can also encourage dependence and reluctance to provide data without payment in future for similar or other project-related activities.

Even with the above survey, it is recommended that a long term monitoring of fisheries should be established as soon as possible. In addition, the Project must decide how to move forward with practical fisheries-related interventions that benefit the widest number of APs, who despite changes to the environment, still show high dependence on living aquatic resources as part of their livelihood strategies.

4.7 **Fish conservation education and village rules**

Deleted: and over one third had to be discarded

Deleted: <#>The survey has provided a snap shot for some variables over one full year (mid-2002 to mid-2003). However, it cannot provide any data for previous years and for variables that were not included in the form. ¶
<#>No provision was made for on-going, long-term monitoring of fisheries, even at a superficial level. Large concentrations of financial and human resources to provide data for a single year is a weakness of the adopted approach.¶

Deleted: As a result of the long delay in data processing and [redacted] finalizing the report, time has been lost that could have been profitably used for long term monitoring activities. It is clear from discussions with the project management that it has been a somewhat frustrating process for staff and management having to wait for so long for feedback from the survey. However, now

Deleted: The Revised Logframe Summary 2003 (November 2003 Draft) suggested that there were a number of immediate fishery mitigation measures [redacted] could implement that required further investigation. One of these was: "Support village committees in

Figure 4.1. Literal marginalization. This is a screenshot of the edited draft document showing deletions, insertions, and comments made by company managers. Note that the strongest criticisms have been removed; many of Steven's interventions are streamlined or toned down, and critical detail is eliminated such as reference to published scholarship and the World Commission on Dams. (Redactions are mine to remove identifying information.)

living conditions. The company might cease its sustainability efforts if it was unduly criticized; this worry defined the consultants' ethical debates.

Obligations to Environment: Writing and Speaking

The document's stated objective of providing practical advice to the hydropower company's environmental staff was an achievement won by Douglas, the rural development specialist, whose experience gave him influence over the otherwise disorganized and dramatic consultancy. Both the fisheries specialist, Steven, and the social development expert, Umporn, offered strident critiques of the project that were not directly relevant to the operational work of the environmental staff. But Douglas ultimately convinced them that overt criticism would erode the company's financial support for environmental activities while scuttling the noteworthy enthusiasm of the environmental technicians, who had clearly distinguished themselves in working hard to put the environmental program in place. "There are many ways to do a report like this," he told me,

> You can come in and assume everything is wrong. You can run it like a snot-nosed audit. You can act like they're hiding things . . . I keep coming back to 'project empathy'—can I do it better? If not, then I'll keep my mouth shut. If there are constraints it's not their fault. . . . You can't come in like a criminal investigation—it won't ever do any good. I'm not expecting perfection. The program has lots of serious problems, but still I'm very impressed. After the first couple days here I saw that the project was good but needed lots of small changes.

Such an approach differentiates between judging whether the project overall has successfully fixed the problems with the dam or, on the other hand, providing practical advice for improving the environmental activities underway. But Douglas's comments also raise important questions about the positionality of their work and their relationships with the environmental managers, staff, and technicians whose work they are meant to evaluate. These questions of relationship, duration, and place dominated the ethical debate, and such an ethics—not merely a kind of abstract moral reflection—took form in the turbulent wake of the dam's construction.

The consultants' relations with the environmental staff formed the primary basis of their ethical deliberations. The staff of the sustainability operations comprised, at the time, about ten full-time Lao technicians, the manager Bouali, and an Anglophone expatriate by the name of John who had developed the overarching environmental program but who formally worked in an advisory role. Bouali, the unflappable, quiet manager of the environmental division, and I met in the company of others, and he took it in stride that Richard, his boss, had let me come observe their work. Seconded from the government electricity utility, he had the studied air of a bureaucrat who believes in his core that information, let alone more reflective rumination, should only be shared if absolutely necessary. Bouali kept things organized and running, while it was the energy of the motivated field staff that drove the operation.

The verbal negotiation surrounding the writing process was deeply involved, and the written artifact was secondary to detailed practical discussions with environmental staff during the consultancy. This verbal process constituted the bulk of the work, in which the consultants spent considerable time in active dialogue with project staff trying to understand the substance of the mitigation approach and provide workable advice on ways to grapple with the most significant operational problems. In turn, this meant that the final document was more like a summary report describing more detailed conversations that had already taken place.

"My interest isn't in technical perfection, and the other two think I'm too slack, too easy," said Douglas, the rural development expert. "I see the problems, but I have to put myself in the shoes of the staff—could I do better?" Gesturing upwards with her index finger, Umporn, the Thai social development specialist, insisted that she was not writing the report for people "up there," such as IRN or company managers: "The staff is very motivated here. We used our time for that. Make it practical, easy to pick up for the staff. When I delayed my time in Vientiane for two days [when deciding whether to quit the consultancy], Bouali called and said 'Why?! I don't know anything of these politics! Just come here so I can improve my work!'" She continued, "I feel 'Ugh! [holds hand to heart]. So I came back."

Douglas echoed a sentiment that I found common among development professionals, especially those involved in technical disciplines, that the technical issues in development were in actuality complex social and institutional problems. Watershed problems were usually caught up in land tenure or other issues, and he emphasized the quality of interpersonal interactions as key to his work. Douglas cited an interest in "how much they're doing different—their hardware and software—they're trying something unique." Umporn was similarly enthused about the breadth of the company's activities, although much more critical about exacerbated inequality, the exclusion of women from major livelihood initiatives, and the failure to fix the main problems with the dam. But Steven's criticisms provoked a very different reaction from project staff, curtailing working relationships and limiting the extent to which his critiques could be heard. This was especially true in raising criticisms that, from a management approach, could not be acted on. But it also had a flavor of imposing his own preexisting terms of judgment, whereas Douglas and Umporn more successfully worked to negotiate the criteria of performance evaluation through the working process with the environmental staff.

I encountered John, the advisor who had developed the environmental program, in the dining hall, fuming over a preliminary report Steven had submitted. The ten or so typed pages were folded lengthwise in his hand. Douglas tried to reassure him. "Steven can't help writing that way, John, that's just how he sees the world. You select consultants because you know someone who knows someone who says, hey, that guy's not so bad. You never know what they'll write—you've got to distance yourself from what they say if it makes you so upset."

John had scribbled comments in the margin here and there. "Bouali came in while I was reading this yesterday," John said. "And he looked at me and said, '*jai yen, jai yen* John'—it means 'cool heart, cool heart—calm yourself'—because I was so upset my blood pressure was through the roof and my ears were turning red."

"He writes all this stuff," John continued, "Da-ta-da-ta-da. Then he writes in bold capital letters, several font sizes larger: BUT—BUT this is what you should do about it. He'll write something like, the fish ponds were stocked with fish BUT the fish were only used for important guests! . . . It's his tone that's a problem."

Douglas and John were leaving as Steven came in to the room. Uncharacteristically restrained, John told him he would just make a few comments, three or four lines, nothing major. "It's nothing to worry about," he said. "Anyway, you write what you write." He looked exhausted as he left the room, but it was Steven's input that became progressively marginalized.

By contrast, Douglas and Umporn went out of their way to work through their recommendations with the environmental management staff, working closely with Bouali, John, and individual technicians, and their approach to working with people was convincing on its own terms. Douglas insisted on recognizing the ambiguity with which assessments are taken up by management. "Project managers get lots of bad information—consultants say all kinds of silly things and it's very hard for them to know what to do with it."

"Before making any kind of recommendation I talk with the staff to see if they understand what I'm trying to say, to see if they can respond. They should have every opportunity to tell me why I'm wrong."

This process implies a striking ambiguity in the kinds of assessments made: they were not based on static criteria but were situationally enacted.[7] While the company managers intervened very little in how the consultants did their work, Douglas's role served to reign in criticisms the other two consultants were prepared to make. But more important were the ways in which criteria for assessing the program were continually opened for negotiation within this process of establishing limits. It is in this sense that one can speak of their ethics as part of a process of evaluation in which the rules are not pregiven but are constantly opened to renegotiation and modification. The consultants were explicit about what obligations to the changing environment might take precedence. Let me turn to a concrete example.

In the ways he worked with project staff, Douglas was impressive to watch. He asked sharp questions that were generous to the staff while probing of the kinds of difficulties—and hence skills—with which they worked. Much of the time I was with him he spent talking to a small number of the Lao environmental technicians, often with a translator. Suchada, a dynamic, charismatic ex-monk who worked on protein-substitution and wet rice projects, told us about his efforts to resolve some of the issues around a still-active, colonial-era tin mine whose tailings were polluting the water supply of some of the project villages. Suchada had gone to the district government to try to get them to intervene. "You are able to go there and tell the district, we have a problem—and then it's fixed? How does that work?" Douglas pried, trying to understand their relationship to the district. Douglas's questions persisted until the project's relation with the district government became clearer.

Over a bowl of noodles for lunch, Douglas continued to query Suchada by asking about marginal cases for farming debt and credit extension. The project is supposed to be so that everyone participates. But how do villagers participate if they do not have any land?

Suchada: In the dry season, even if the land is owned by one person, for the season it is divided up according to available labor, so that one laboring adult can farm one rai [0.4 acre]. In the wet season, really everyone has some land.

Douglas: OK, take credit. A farmer borrows for seed and fertilizer, but even after a good harvest he must pay his old debt to his uncle, pay for medical bills or for borrowed rice—what happens to the village fund?

Suchada: Yes, this makes it difficult. Usually farmers had to borrow rice four or five years ago [i.e., shortly after the dam began operations], or this year they ran out of food and had to sell green rice, or maybe old debt to an uncle—usually it's these sorts of things. We can give him a second chance. We talk with the village and with the Village Development Committee [set up by the project to manage participation]. It's village money, not the farmer's money, and not our money. We talk with them all together. Sometimes they can pay back little by little, like if they have a government salary. Or they pay back only the capital but don't want to pay interest.

Answering this barrage of questions, Suchada could hardly eat his soup.

What is relevant to my discussion are the criteria by which fundamental relationships—the question of villagers' dispossession and the obligations of the hydropower firm in composing new life possibilities—are integrated into a critical assessment. The problem of land access was acutely felt by many farmers, in ways the final report alluded to but failed to make explicit. And Suchada acknowledged that at least some of farmers' debt was accrued after the dam began operations in ways that raise questions about the firm's compensation liabilities. Issues that were edited out of the final report in fact have a very specific relation to what went into the report, and this in turn was negotiated in the verbal processes through which the document was produced.

This enactment of negotiated value in the gap between writing and speaking, in part designed to protect the company's operations while confronting the problems it faced, foregrounded Suchada's obligations to his work, to the problems of villagers and the dynamics of a complex river system. A prominent underlying tension was the remarkable amount of work required to have any kind of substantial effect on people's livelihoods, a pathos juxtaposed with the minimal infrastructural resources for addressing such site-specific problems. If Douglas gave company technicians a pass for not addressing the hardest problems, it was from a recognition of the difficulty of acting that Suchida and other technicians faced. At the same time, the kinetic relation of active negotiation affirms Suchada's work with other experts, not villagers' problems per se. In turn, whereas Steven's mode of critique served to progressively isolate him from the team, Douglas used the writing of the report for the active construction of technical relationships in such a way that its major critical interventions were already negotiated in the writing of the document. This raises the issue of professionalization, for report writing itself is the hallmark of the consultant's professional output.

Terms of Reference

Douglas was the one primarily worried that Steven could not complete his duties. "Sometimes a consultant who is unable to complete the terms of reference will submit a trip report where he describes what he did, what he thinks, who he talked to. In that you can include whatever you think about everything." As for himself and Umporn, he said, "We should be alright. She'll probably write an annex describing some things that aren't in the Terms of Reference." The criteria of professionalism for which he advocated, normally not so visible in a routine consultancy, here came into view around three distinct dimensions of the report: the question of audience, whether the report would be focused on environmental impacts or on project activities and their achievements, and the extent of its technical criticism.

Toward the end of the consultancy, Steven only talked about two things, and he mostly did not reflect on his practice. Those two things were a fisheries study that was supposed to have been completed before he arrived, and the dismal state of the fisheries as an unavoidable effect of large hydropower dams. While himself under fire for being unprofessional, he challenged the professionalism of the author of the incomplete fisheries study. "These conclusions are really embarrassing," he said of a faxed, draft copy of the unseen report's conclusions and recommendations. "He has really sold out his profession. Show me a recommendation—show me a single recommendation in there!"

The predominant issues for Steven were not how to improve the technical practices surrounding the dam, but to document hydropower's mostly inescapable impacts on local conditions for living. For his colleagues on the team, his criticisms seemed too instrumental in using this particular situation to make a broader case against dams in general. One could say that his critique was technical in content, since he addressed the dam's effects as an expert, but not technical in form since it was not primarily oriented toward expert advice in an organizational context. When I asked Steven why he had decided not to quit when he had the chance, I expected a reasoned explanation about his interest in the activists' work and his commitment to riparian ecologies and villagers' livelihood issues. Things were going badly, however, and his answer was far more ambivalent. In fact, he did not provide an explicit personal reason for continuing, mentioning only that the activist group had suggested they would like for him to stay and that he saw "little point in stopping now."

Much of this tension played out in debates about the document's scope. Douglas was persistent that the audience of the report should be only the environmental management staff, as an aid to their work—a position clearly favored by the company's management: "I asked Richard, 'Is that where you want us to go? Is it too technical, or maybe not enough?' Really the issue was if the report was for the world or for THPC. These would be very different reports. You don't need much detail if it's for yourself. Focus on the existing mitigation activities, see how they all fit together, what shortcomings or adjustments can be made, and always ask the question whether they're doing the best they can. It's not complicated." Of course it *was* complicated, since this enforced, narrow technical criticism excluded some of the most controversial socioecological effects of the dam. Since Steven's job was to assess what the sustainability efforts had done to improve fisheries to date, he was left with not much to do; since his comments were meant to be restricted to practical recommendations, there was not much he could say. Overall, the problems with the fisheries would not be immediately addressed by any technical feat shy of restricting dam operations. Compounded by the very clear negative impact of the dam on the riparian ecology, from his perspective, there was little positive to affirm and his critical voice was progressively marginalized during the consultancy.

Douglas continued: "I'm quite positive the other two are positive on the social work. But Steven is a globalist looking at the ecological degradation of the fisheries, and he feels quite responsible. His TOR [terms of reference] is to save the world—mine no longer is. He wants to do an assessment of the whole shooting match—to document every last thing—whereas our job is to see how the different parts of the project are working. He wants to go into many subjects that—so far as I can see—are not in the TOR but his own interests. He's very angry and he's taken it upon himself to—. . . ." He cut himself short and failed to finish the thought. But later turned to a personal reflection on anger: "I used to be angry. I've shut down projects before that I thought were really bad, but generally if I can't do it better or can't make positive recommendations then I don't have much to say."

Not having much to say could be viewed as an extreme limit of what Douglas was arguing for—a kind of strictly enforced positivity. But it could also be viewed as a restatement of the problem of the pitch of the anthropogenic, its tenor and its wager, or the way in which the tone of writing demonstrates expectations, obligations, and commitments worked out in the process of writing itself. The forms of professionalism relieve some of the burden of working on the front lines of the anthropogenic. It is through a career of professional development that the disposition toward technical criticism is adopted—an interpretation confirmed by Douglas's reference to his past. Furthermore, as Umporn's considered reflection shows, "not having much to say" raises the question of to whom one might be able to address criticism.

Audience

Trained in participatory approaches to natural resource management and working as a researcher with a noted Australian research center, Umporn came to the project out of an interest in the experimental process with the activists and a desire to see hydropower improve. Like Steven, when the process threatened to fall apart she seriously questioned her involvement with a hydropower firm. "When I found out about it [the experimental process]

I thought, 'Oh, interesting problem!' I wanted to participate to see how it turned out." She cited in particular a dislike for activists who simplified what happened in villages, who neglected that villagers often lived hard lives and, therefore, had legitimate claims on development. At the same time, hydropower projects exacerbated these situations even when they held potential to provide benefits.

When I asked her whether John, the environmental manager, had pressured her to write a certain kind of report, she emphasized the integrity and motivation of the staff. "Maybe 50% forced," she said. "Anyway, we could complain if we wanted to. It is a problem of time. Not 'power' forced, instead force ourselves to do good because the good work of the environmental staff here. If they are not so good [we might not try very hard]—so forced is kind of push and pull. If we spend too much time [on one thing], another group won't benefit so much." Their effort in the sustainability work convinced Umporn to adopt a series of professional obligations. Staff motivation was therefore critical to holding together the performance assessment team. Was the sustainability work in the villages any good? It was good enough to hold the consultants together, to forge out of their assessment a technical engagement with the details of environmental management work.

Yet Umporn's interest in the process was much broader than providing narrow technical recommendations. "I'm more interested in what the people think about the dam, how they adapt to it. What do people think? Why do they accept the project? They [the hydropower firm] have limited staff and time, and in many villages they are not too effective. Right now, the villagers are told to do this and this, and the villagers follow. In the first 1½ weeks, we tried to get a picture or feeling of what is going on here. I didn't expect good people. But then, now that we're almost finished, we go back to the TOR and just make comments on the activities." Umporn very clearly made a deliberate decision that her best input would take the form of careful advice that improved, to the degree possible, the immediate work of the environmental managers. Her choice to focus on technical criticism displaced a broader set of her own concerns about the bigger problems that remained. Yet her approach acknowledged the fact that activists would not be the ones doing practical work to make villagers' lives better, and that, should the firm fall back on its environmental work, the government would hardly step in to take over those tasks. Technical criticism was the only responsible choice.

"The last thing a consultant should do," Douglas tells me, "is write what his employers want. You can write from the left or from the right, but you have to be consistent, you have to say what you think. Otherwise pretty soon no one wants to hire you."

"Is it a question of professionalism?," I ask. "Or simply that no one can trust you?"

"Yes, exactly, those are the same thing. Otherwise people who've hired you might as well have written the report themselves. It would be no different. If you get any two consultants—it doesn't matter what project—you'll get a different slant. If you're too quick to bow to the client then no one wants you."

To whom can one voice criticism? Umporn worked closely with Douglas to pinpoint problems so that her input would be of better use to the project's technical staff, effectively engaging Douglas as a mentor and in a process of clarifying her commitments. She made the conscious decision to write for the environmental staff and managers and not for activists or others trying to gauge the project's limitations from afar. Quite concretely, that implied a commitment to the environmental staff, to Bouali, Suchada, and others working in the long term with villagers. It was their motivation that pushed her toward a different critical frame, and I read her decision as a commitment to a certain kind of relation in which it is possible for her critique, and therefore her knowledge, to be heard.

Conclusion

The many tensions that crisscross the document should not be read simply as a hydropower firm trying to write its own report. There is no doubt that the company had a clear interest in influencing the content of the document while creating the appearance of its independent basis. Furthermore, I have not shied away from showing the process of eliminating specific content and toning down the text. Yet the report is rich with detailed criticism, and with a little careful reading it shows that nearly all the most important environmental problems had not been successfully addressed by the sustainability program. Perhaps more importantly, the highly process-based approach of working out the terms of evaluation through negotiated professional and expert relationships served to foreground the situational complexity of sustainability practice.

Rejecting the tactical demands of both the activist group and the hydropower manager, these consultants collectively and individually were concerned with what Michel Foucault (1990) calls a concern for truth. That truth included not only questions of fact but, more pertinently, questions pertaining to the value of these facts and the worth of criticism. Their ethics concerned their relation as experts to other experts and therefore involved directed exploration of the obligations and demands of the hydropower company's embedded concern for the anthropogenic river. This required not the application of a set of stabilized, external standards that would supersede or judge another from a distance, but rather took form out of a concern with the challenges and uncertainties facing the project environmental staff themselves. This concern with others' uncertainties is one possible figuration of an ethics of the anthropogenic that expressly considers the value of environmental knowledge. Furthermore, while there are many practices of expertise when it comes to anthropogenic environments, one can also identify here an ethics of expertise in which it must work as a kind of criticism capable of being heard. This form of directed criticism necessarily places expertise at odds with a general political critique and forces attention on the "clinical" dimensions of environmental knowledge. Yet it also requires that technical criticism not simply limit itself to improving the narrow concerns of technicians. Technical criticism must be tempered so it is capable of being heard, but so too it must articulate the cares of those who are not included in the process.

The mode of the anthropogenic took form in terms of an experiment to establish a degree of independence in consultants' work, and the very active debates about the role and practical implications of environmental expertise. I have argued that the ethical dimensions of this situated endeavor revolved around the experimental basis of formulating articulated criticism and judgment; in the routine vetting of advice, criticisms, and recommendations between expatriate short-term consultants and the Lao environmental staff whose long-term commitment was directly relevant to village interventions; in the problematic of to whom critique may be addressed and in what terms; and finally in the consistency of evaluative terms that develop over time and through negotiated process. I have also argued that the ethical takes on special significance not in the absence of power relations, but through negotiating compromised situations shot through with questions of power, inequality, and dispossession. As Faubion (2011, 125) suggests, the consultants are acutely aware that what they do and think today means more than what they can possibly anticipate.

An understanding of ecology as a problematization defines the space for an ethics of the anthropogenic. Viewing ecological knowledge as a problem domain in relation to an event rather than a program of positive knowledge or discursive power does not mean neglecting those dimensions of power/knowledge, but rather attending to broader discursive and nondiscursive factors with which they attempt to engage. Problematization marks a shift in Foucault's thought away from the determining qualities of discourse toward the array of discursive and nondiscursive practices through which questions and answers are intelligible and valuable. This can be opened up further by scrutinizing not only nondiscursive practices, but also nondiscursive relations.

While for Foucault problematization was a way to explore practices of thinking in their relation to historically singular forms of experience, there is also a way to use the term in service of an ontological pluralism.[8] In fact, while the methodological emphasis on the term "problematization" does not appear in its full form until very late (see Rabinow 1997), a decisive early moment seems to be Foucault's 1970 engagement with Gilles Deleuze's *The Logic of Sense*, of which he wrote a mind-bending and at times ecstatic review (Foucault 1998). Deleuze's notion of the problematic, from *Logic of Sense*, directly addresses the status of the event: "The mode of the event is the problematic," Deleuze writes. "One must not say that there are problematic events, but that events bear exclusively on problems and define their conditions" (1990, 54). "The event by itself is problematic and problematizing" (56). This chapter therefore helps show the extent to which debates and political contestation about ecology function neither as a totalizing mode of power/knowledge-cum-resistance that defines capacities for thought, nor as a pure struggle in which politics replicates a form of warfare, but as a problematization that maintains a relation of openness to an "historical ontology" of the human (Foucault 1997c, 315). For Foucault, the ethical demarcates a critical task of philosophy that is in fact anthropogenic. "The philosophical ethos appropriate to the critical ontology of ourselves [is] a historico-practical test of the limits we may go beyond, and thus as work carried out by ourselves upon ourselves as free beings" (1997c, 316). Furthermore, "this historico-critical attitude must also be an experimental one" (ibid.). The anthropogenic maintains a fundamental ethical dimension not because environment is a moral issue but because ecology is necessary to the critical ontology of anthropos in the contemporary.

When Steven and Umporn—and Douglas to a lesser extent—committed to continuing their contract and producing a completed report, they placed considerable emphasis on the motivation and energetic commitment of Bouali, Suchada, and the rest of the environmental management staff. In chapter 5, I look more closely at their work in order to understand better the tensions between uncertain environmental knowledge and tenuous sustainability interventions. The experimental ethos of the environmental program was described repeatedly as one capable of incorporating the flexibility of trying out many ideas to see what worked, rather than attempting to fully understand all the variables before committing to a course of action. Thus their environmental work was implicated in a specifically neoliberal program of action that, one could argue, thrives on uncertainty. But there is another dimension to their activity as well, and that is the question of the Lao environmental staff's "motivation." It turns out that their primary motivation was numerical. Inspired by the quantitative achievement targets in the logical framework device, an instrument developed by management theorist Peter Drucker, they had eagerly embarked on achieving as many environmental targets as possible. Given that context, chapter 5 explores the full extent of the anthropogenic river as sustainability practice comes face-to-face with those people their work was meant to help.

Interlude

Interview Notes (Lightly Edited)

Note: Quite high up in the watershed, these villages were outside the hydropower management area and were not discernably influenced by the dam's impacts or the sustainability management program. I provide the notes to give a sense of the contours of village livelihoods.

Ban Hat Sai Kham Households Resettled from Ban Fan

Interviews October 20, 2004, morning. Many people were in fields today; talked mostly with young and elderly, except for one man about forty-five years old. Tried to interview Hmong households (hhs) also, but literally no men were still around, except for one man who was busy drinking and not interested in talking to us. Hmong women were not interested to talk to us. Working with Singphet (research assistant).

Sixteen hhs still in Ban Fan, by one estimate; about eighteen have moved here. But many people sleep at fields during the harvest. Villagers are required to move—all of them—only some have not moved yet. Rest of village may move this year—already have house plots here. Fields are one or one-and-a-half hour(s) walk from new village.

Young man's parents are at hill rice fields with rest of family; he is seventeen years old and sick today. His family came from Ban Fan three years ago. Two sisters return from school while we're talking. There are eight in his family—mother and father, twenty-one-year-old brother and his wife, who looks about sixteen (shows up during interview), himself, and three younger siblings. Normally they all sleep here (quite small house). He has an elder sister with husband and child who live in another hh here. He went to school only for primary one; then stopped to work in fields around age thirteen or fourteen. Younger sisters are more advanced in school; brother is in primary two. Does not know ethnic group; says upon asking, yes, maybe it is Nyo.

Their hill rice is upstream on the Gnouang River, along Houay Khom. Before their fields were in Ban Fan. Moved because of the district government's village consolidation policy. They were poorer in Ban Fan, although at first he tells me opposite. The rice there was fine, but they did not have any money. It was there that he went to school, but the problem was the teacher, who was responsible for many villages—he taught in many places, therefore only a little in each.

Here they have enough rice to eat; they do not need to buy rice or ask from neighbors. Five *hoy* this year, equivalent to about 1.2 tons. Units used are *hoy*, *ķam*, *fat*: Four *ķam* is one *fat*; *hoy* is 100 *ķam*. *Kam* is a "handful," unthreshed grains on short stalks (as they are often dried), about 2.5 kilograms of unmilled rice. A *fat* is therefore about equal to ten kilograms of unmilled rice. Properly speaking, they are all volume measurements. They are able to eat rice with every meal. Also eat a lot of forest products—mostly rats, bamboo shoots, gathered vegetables, fish (two to three kilograms/hh/week). They spend about two hours a day gathering in the forest; everyone goes. He makes rat traps with bamboo.

This is their second year planting on same soil. Reusing the same plot is called *hai loķ*, or "pull-out hill rice" gardens (i.e., rather than cut fallow forest). Last year they harvested 7 *hoy*—about 1.7 tons. They usually rotate fields after two years. Next year will have fields in Houay Kadae; their house will stay here. First year after move had hill rice here, then moved for two years in same place. Can move after one year if the soil is no good, or stay up to three years if good. Government never asked them to stop hill rice.

Family has no livestock; never had any large livestock or pigs. Had over twenty chickens but they all died from disease. They have a cat, which is a mouser, bought for 10,000 lak (US$0.95) from a Vietnamese trader.

Family is heavily dependent on wages from casual labor. Two people, the father and mother, work about three days each per season to help others with harvest to earn 15,000 lak (i.e., approximately US$1.42) each person per day. Elder brother cuts fencing posts to sell in Ban Hat Sai Kham for chili gardens, about 100 posts bring 30,000 lak (US$2.85), about a day's work. Does this every day. Posts are about three centimeters in diameter. This brings in about 200,000 lak (US$19) per week. No one in family works for Hmong due to mistrust. He works clearing gardens often—about 170,000 lak (US$16) per area, does not know how wide. Very hard work, takes more than ten days to clear.

Opium was prevalent before, perhaps still now. Farming income was 100,000–1,000,000 lak per hh. His father smokes every day at a cost of 15,000 lak (US$1.42) per day (!). Does not know about marijuana.

At one hh talked with elderly blind woman, perhaps late sixties (claimed to be ninety—that mythical "old age" age). She came here because her granddaughter was sick, came with son's family moved here after everyone else came from Ban Fan. Came to be near granddaughter while she was treated. Says family was happy to come here—this is a focal zone (*khet jut soum*) village and has a clinic. Did not leave Ban Fan earlier—in other words, with everyone else—because they wanted to wait and see about the situation. Sold their large livestock to pay for granddaughter's sickness. Nyo ethnicity.

Man and woman over forty. Nyo. Says hill rice prohibited in Ban Fan next year, or rather prohibited to cut trees only. Government land titling here. Actually this is just for gardens, which people get thirty to fifty square meters. Hill rice is not yet measured. They can grow rice wherever, but they cannot cut trees—they can only use fallow fields. Seems like he is just talking about land use allocation—maybe he is not clear on the policy. Planted two hectares hill rice last year; should yield enough for two years. Has small family—five people, two under fifteen years old. About 3.8 tons from two hectares last year. Uses same *fat/hoy/kam* measurement system. Not growing hill rice this year. It turns out that his daughter has died (mentioned by elderly blind woman, his mother). Did not plant this year because she died early wet season, could not do the work. No real income sources—150,000 lak (US$14.30) annual income from chili and lettuce—also stopped chili because daughter died—no forest products, no hunting, no casual labor.

Says people in this area used to earn a lot from marijuana, stopped six to seven years ago. He seems to be including himself when he talks about it. Very quickly interrupted by a nosy man named Phon wearing army clothes that are too big for him, who was apparently listening to us from around the side of the house. Phon tells me "everyone here is just simple people, we don't know how people live in international countries (*phathet sakon*). We just live by our gardens and our rice, don't even know about growing cash crops, or how to sell anything." Then he turns to my research assistant, who is carrying a digital camera, and asks him, "Is that a tape recorder?"

Chapter 5

Anthropogenic Rivers

In the fast-running streams and rapids of Laos's mountainous waterways, hydroelectricity has long been a feature of riparian lives. Small dynamos, Vietnamese-manufactured generators about the diameter of a dinner plate with a meter-long shaft and propeller, announce the approach to villages if one is hiking along the paths that inevitably track the banks of settled waterways. Thin electric wires rise up from the water's rapids, propped up on bamboo poles, stretched across rice fields, their wires' ends spliced together by hand. A dynamo can power a couple of lightbulbs or wired in parallel, a TV set and DVD player. Such "pico-power" units (Smits and Bush 2010) are rated at several hundred watts, but the resistance of the transmission wires saps their voltage, and the generators inevitably have been repeatedly repaired with their internal wiring carefully rewound by hand. In rapids, dynamos are mounted on wooden frames with rocks arranged around the propeller to channel and accelerate the flow. In smaller streams, villagers frequently build a wooden or rock funnel to concentrate the water's energy.

Design matters in these configurations, and arrangements of hydro-gravitational flow are one of many domains of acute skill through which living makes sense according to distinct axes of value. Gravitational flow, the flow of electrons, and the sociality of electricity—very often for communal viewing of Vietnamese television or Thai karaoke videos—are brought into alignment, if only for a time. There are, moreover, dozens of ways in which people in these villages block, manipulate, channel, and divert these riparian flows. On smaller streams, weirs are frequent ways to cultivate aquatic habitat. Irrigation is less common—it plays a minor role in the technical repertoire around these mountain streams. But it is foolish to imagine that these rivers were not anthropogenic prior to the construction of large dams. These technical arrangements are not opposed to ecology, rather they are specific intensifications and affirmations of certain aspects of that ecology. Insofar as the technological artifacts are concerned, the difference between pico-hydro and the Theun-Hinboun Dam is a difference of scale. One is two hundred watts, the other is two hundred million watts. Six orders of magnitude demonstrate a "passion for difference" through which the gravitational flow of water is wired into the histories of Laos's late industrial rivers (Moore 1994; Boyer 2015).

By staying close to the modes of expertise at work in Lao hydropower, so far I have asked readers to dwell or tarry with the possibilities of a late industrial environment in which "technological artifacts temporalize, opening us to a future that we cannot fully appreciate even as they render us subject to a past not of our own making" (Braun and Whatmore 2010, xxi). In this chapter, this generative temporality is framed around a dominant material-semiotic form, the logical framework management matrix, that orchestrated bubbles of self-motivation and intensification, among the environmental technicians employed by the company, and certainly many villagers, too. Characteristic of neoliberal forms of self-management, these bubbles of motivation form a kind of "entrepreneurial life" for environmental technicians and some affected villagers. I address the ecological dimensions of living for villagers along these two rivers, in the riparian interplay between wet and dry that has been fundamentally changed by the dam itself. Managerial modes of expertise and entrepreneurship rub up against multiple temporalities of the river and its familiarities, and I explore languages that emerge when the limits of aspiration are only too apparent, when people do not know what to do, or they understand little about the environmental relations in which they live.

Elizabeth Povinelli (2011) offers a language for working within such a temporality. She describes the "ordinary, chronic, and cruddy" forms of living and dying that constitute a biopolitical mode of abandonment and "letting die." Here, letting die realizes a kind of effacement in which forms of living are unmoored from the conditions that make them possible. It is not literal death—no persons expired due to the circumstances of the dam, to my knowledge. But living is not bare life, like a biological minima devoid of the pluralities of a full existence, or even life itself, which is a genetic problem (Rose 2007). It is rather the rich, expansive existence in which people are formed in mutual relation with the historical and meaningful situations in which they live, and it is this expansive existence that ceases to be possible in the shadow of hydropower dams. Thus Lauren Berlant allows us to see entrepreneurial life as a kind of "cruel optimism" amid the "speeds with which capitalist activity destroys its environments while at the same time it makes living possible and produces contexts of thriving, merely living and wearing out for the people making their life within them" (2011, 115). The production of uncertainty in these riparian ecologies disables specific conditions of living and renders certain worlds unintelligible, but it is hard to argue that it unilaterally destroys life.

Hence, while this biopolitical assessment is necessary, the anthropogenic exceeds the biopolitical frame. Throughout the book, I have used the term "anthropogenic" in the double sense of pervasive ecological change originating in the activity of humans—the conventional sense—and the permanent emergence in which the human and its capacities are a work in progress. If we think of the human not as a predefined biological (or moral) entity but as an ill-defined amalgam of perpetually unfinished business rich with irreducible capacities and obligations, then the anthropological question shifts to address the terms of the ecological contemporary. Human capacities for bearing and inflicting harm, including ecological harm, have radically expanded over the past several centuries. Anthropos today must take these capacities as central to the task of anthropology. These capacities, like other more noted achievements, such as knowledge of the physical universe or the understanding of the molecular conditions of life, have pushed the limits of the human. While this research is not designed to address suffering per se, because my argument raises questions of value and normativity it is important to underscore that people do not merely experience harm, they also bear it, reject it, and find ways to work around it—in short, people judge harm in relation to their own capacities, and they establish new norms regarding the harms they are capable of bearing or rejecting.

Hence, the anthropogenic has a creative element in which people speculate on possibility and actively deliberate on the value of their activity. Likewise, damaged ecosystems result not merely in harmed bodies, but in persons capable of bearing harm, which is a powerful capacity in its own right. Ecological harm is frequently subtle, for it undermines diffuse, tacit relations in ways that do not necessarily form part of everyday explicit recognition and discussion and therefore places demands on language and thought. Finally, the anthropogenic has an exploratory and subjected element that comes from being situated within novel obligations in which ecology, economy, and politics are frequently indistinguishable. At the limit, the possibility of living fuses power and life, knowledge and ecology. These rivers have always been anthropogenic, not because the anthropogenic is an Edenic caesura, a genetic principle or a tale of genesis, but because it is a permanent condition of emergence that late industrial environments have radically intensified. I address the biopolitical modality of the sustainability enclave before turning to the interplay of ecology, economy, and politics.

Entrepreneurial Life

The dam itself was far out of the way of most of the deliberations that have characterized this ethnography so far—its traces were pervasive, familiar, and independent of the immediate presence of the thing itself. A man named Somvang first brought me to the dam. We had taken a detour after visiting some of his fields further up the headpond where he checked on his crops and explained different styles of rice cropping. In a nerve-wracking display of his familiarity with the dam's immensity, he motored his metal canoe just above the dam's spillway as mist from the powerful cascade of water billowed up around us. We climbed above the far side of the dam and watched the churning, turbulent flow in the river below (see figure 0.1). Come the dry season, the flow would be clenched to a trickle.

A vice chief of a village near the hydropower dam, Somvang was an agricultural entrepreneur and the virtual image of the successful Lao farmer for whom the environmental management program was designed. In his early forties, he was a de facto spokesperson for the energetic trialing of activities by the hydropower company and the implementation of district development policy—at times, the two overlapping regimes were virtually indistinguishable. "In Vietnam there is very little land," he told me, repeating a classic party line about rural development, "but people make good use of it. Laos has used up its forest but is still poor." He had been a Lao People's Revolutionary Party member since he was fourteen, my research assistant later told me, and his rendering of district and company policies was confident and unironic. His village was one of several in which the company's environmental program had been especially intensive.

Experimentation is often an exercise in modelling; village livelihoods are no exception. Take subsidized latrines, for instance. A household would pay into the program, approximately US$4, according to Somvang; in turn the company provided gravel, sand, cement, metal flashing, polyvinyl chloride piping, buckets, a toilet, and the septic form for casting the cement. Finally, the household would then pay into their personal microcredit account, ostensibly to "repay" the company for the materials but actually providing nominal household savings. This is one sort of activity design undertaken by the technical staff of the company's environmental program, with the details worked out in negotiation with village committees. Several men had gone further by using the company's casting form to make extra septic tanks for sale. In effect exchanging labor for capital investment, with nominal capital repayment going into personal savings, latrine construction exemplified the "development of the self" (*patthana tua eng*) ethos of the environmental management program.

Major interventions in Somvang's village:

- Rice intensification
- Reform of land use policy
- Agricultural extension
- Nursery for commercial gardens
- Maize grown under contract with commercial buyers
- Garden irrigation
- Public water supply
- Subsidized latrines
- Village microcredit system
- Health training
- Experimental fish ponds

In another village below the power plant, the company ran a model "farm" (*fam*, a direct borrow from the English), complete with a North American–style farmhouse, which trialed commercial pig, poultry, and rabbit production. The farmhouse was actually a holdover from the previous management, which had it built apparently out of a romantic desire for an idyllic rural retreat. Semi-defunct, it had been converted for piloting multistage production of livestock and feedstock. Village pigs are ubiquitous in this part of Laos, but the company wanted key villagers—comparatively well-off men with extra time and resources, including Somvang—to take the lead with producing commercial pork for market. The animal and its requirements are totally different. Village pigs are small and fatty—small enough to slaughter for household consumption or to distribute among a few neighbors—and they are simply allowed to forage for kitchen scraps. By comparison, the commercial hogs are far too large for local consumption, and a much leaner meat. The program involved subsidizing piglets if villagers invested in constructing pig pens and dedicated garden space for feed production. Aside from the model farm, this village was a particularly energetic hive of activity. Its comparatively new houses articulated along a straight, raised road built by the company on high ground, in part to accommodate newcomers hoping to benefit from local development. The old village sat adjacent, by comparison a swampy, organic mixture of unreconstructed village life and active disinvestment. Once again, the entrepreneurial life played a central role. "The truly poor here are the original inhabitants," an elderly *Kaleng* woman said during an interview, invoking the term for indigenous people (*pheun tham*). Land poor and without children, she had sold her house and now stayed with her sister or other relatives. "We are poor like before the dam," she said. "We have no capital, we don't know how to use it. We just live naturally (*tam thammasat*)."[1]

Pig pens and latrines only go so far in articulating a program of managerial care. In Somvang's village and many of those adjacent to the reservoir, the company had worked closely with district government to implement land use allocation. This involved restricting people's ability to practice shifting hill rice cultivation—their main source of livelihood—by assigning them permanent plots, one hectare per household. They had also invested variously in expensive chemical fertilizers and labor-intensive organic composting to little avail, while in other cases the company facilitated agricultural traders to develop contract farming initiatives for maize or tobacco—providing seed, fertilizer, and credit in exchange for a lien on the farmers' crop.

In several villages, THPC had further initiated an intensification program to try to convert these hill rice plots to wet rice "paddy," for which farmers dug terraces into the thin top soils in a rice-for-labor program. There is no paddy here, I was told by one farmer—only "paddy-style" rice (*paep naa*), invoking an apprehensive term that had come into common usage for these weird terraces without adequate water supply and none of the rich ecological dimensions of wet rice agriculture. "Permanent shifting cultivation" (*hai k̄hong ti*) was another neologism that attempted to grasp (apprehend) a situation of unpredictability in which villagers were required to plant repeatedly on the same plot rather than rotate their fields. In the second year of this experiment, the crops were failing badly. The plants were infected with rust, weeds crowded them out and made inordinate demands on labor, insects took to their roots, and the grains were empty husks with no seed. Farmers in several villages rebelled in an attempt to "quit" the environmental program. Forty-two farmers in Somvang's village had trekked deep into protected forest to cut new fields, only to be fined or arrested by district officials. Another group initiated a petition asking the company for a cash payment and to forego their participation in the program. More difficult to discern was what had motivated farmers to go along with activities that seemed foolhardy in retrospect.

Motivation and personal development were key terms of the program. "After two or three years of not having enough rice," Somvang told me, "people will change their occupation to produce for market." The model project citizen was well aware of the secondary effects of the project's mitigation activities and district land use policy. As he indicated, forcing people into market-oriented production could be construed as the objective of overlapping district and company policies. He appealed to his own investment in pig farming by way of explanation. Invoking the necessity of the market written through neo-Leninist party ideology, it was a bizarrely blunt conjunction of developmental imperatives. Of the 102 households in the village, the man said, only 8 were able to grow enough rice that year. The limits of this entrepreneurial life were on full display.

Before exploring these relations in detail, it is important to assess several contextual elements of power and cultural knowledge. In the section that follows I provide a brief discussion of major touchstones for villagers' understandings of power, nature, and economy through exploration of local dynamics of rule.

Does Sustainability Translate?

The combined effects of the activist campaign and the Anglophone managers worked to create a sustainability enclave in which the Lao state played a minimal role in environmental governance. The functioning of the enclave is conjoined with variegated pragmatics of local rule akin to what Ong (2000) describes as graduated sovereignty characteristic of neoliberal globalization in other parts of Southeast Asia. These overlapping, heterogeneous regimes create a multiplier effect in which party connections, programs of resource extraction, and novel articulations of care combine into ad hoc, contradictory juxtapositions of rule. At the same time, life, nature, and knowledge have a political history in Laos that deserves to be outlined, for it forms the basis for articulations of value and affirmation that do not neatly coincide with the transnational language of sustainability. I therefore take this opportunity to explore local rationalities of power and nature essential to the experience of riparian transformation.

Unsurprisingly, Lao villagers and local government in this region were rarely motivated by the language of sustainability or the political aspirations it may inspire, no matter how much it framed transnational investment and rule for specific sites in Laos. Indeed, the Lao term *phatthana yeun yong* (literally "durable development," with no explicit environmental connotation), while invoked occasionally, seemed largely irrelevant to negotiations between the environmental program and villagers. An overarching spirit throughout this part of the country was one of accelerated transformation in which development had arrived and opportunity was afoot for those who could find ways to grasp it (see Singh 2012; High 2008; High et al. 2009). Yet there were clear tensions between those with an acute "will to improve" and others who found their lives and livelihoods displaced. Furthermore, as with national debates, these discussions tended to affirm or reject a relation to expertise—a thread that runs through the discussion that follows.

In the area surrounding the dam's reservoir, prerogatives of government hinged on a village consolidation program as the central feature to state-sponsored development (see Baird and Shoemaker 2007). The approach was to combine several nearby villages into a single administrative unit identified with one of those villages, frequently whose party links allowed it to curry favor. Ideally, those villages would be physically resettled into the central unit—a process that was vexed by villager recalcitrance, lack of government resources, and open-ended negotiations between villages and district officials. Criticized for years by the international development community, which had previously promoted the policy, this "focal zone" resettlement approach sought to better distribute government services such as health clinics and schools across a sparsely populated landscape while consolidating its control. The program also dovetailed with a land allocation program that sharply reduced farmers' ability to practice rotational swidden, at least in some areas, as part of the party's "Turning Land into Capital" policy effort (Dwyer 2013, Vandergeest 2003, Barney 2014, Baird 2011). Tania Li (1999) describes the importance of intimate "compromises of power" in Indonesia's uplands development, which can minimize the worst consequences of ambitious plans of improvement. Villagers' ability to evade state programs and the difficulty of enforcing development, which might normally mitigate the worst effects of wrong-headed plans, were preempted by the momentum and resources made available by the hydropower company. At the same time, bureaucratic failure often leads to a state intimacy in which the failings of state projects get woven into affective and practical dimensions of living (Herzfeld 1996). Here the multiplicity of rule creates compound effects, contradictory juxtapositions, and divergent possibilities for negotiating the terms of governance and citizenship.

Imagine the stress of village chiefs, who are effectively in the position of implementing a potentially disastrous policy in their own villagers. With the new policy, "it's easier for the government to look after (*beung yaeng*) the villagers," one vice-chief told me; his village was in the process of being consolidated with several others. With my research assistant, we picked apart the meaning of this term, which was unfamiliar to me—"paternalistic surveillance" might be the best translation. "Consolidation is not so easy—the villages are far from each other, and it's hard to make people follow the rules." Repeatedly he affirmed the purpose of the state project—the convenience of the road, the school, the clinic. He was unambiguous that village consolidation meant more administration. He showed me a defunct Soviet power generator, left over from an electrification project in the 1980s. His village had only four houses left—"these are field houses," he insisted, in spite of the obvious fact they were full homes, not the minimal sleeping shelters used during the production season. They sat amid the remaining deconstructed posts of their neighbors' former homes and the characteristic fruit trees of Lao villages. "All of our fields are here, an hour's walk" from the new site. An hour's walk is a long way for a government official.

Everything hinges on this mode of effacement, which constitutes a kind of "local undertone" (Obarrio 2014, 206) of political speech resting on the subtlety of allusion and the unspoken. The interview just referenced was taken outside the hydropower project's sustainability zone, but another village's experience, within the project zone and far from any road, was similar. In this case, consolidation also meant becoming subject to implementation of a land use policy that sharply reduced access to customary lands by assigning each household three hectares for swidden farming. Normally, cropping would rotate between fields on a six- or seven-year cycle, allowing brush to grow, capturing nutrients and suppressing the weeds that inevitably crowd out the rice plants on these thin moraine soils. Three hectares by contrast meant a three-year rotation. For many of the hydropower project villages, the restriction was even worse: a mandate to farm the same hill rice plots every year, supposedly made technically feasible by the sustainability project's attempt to convert hill rice to paddy.

Once again a kind of crypto-politics emerged, in fact the inverse of the "politics of listening" I described in chapter 1, this time calling attention to the limits of expertise. Many families had moved from this village, but two remained in place, arguing that they were "not finished building" their houses in the new village. Surrounded by luscious fruit trees, we sat on the porch of a house solidly built with heavy hardwood planks, a testament to the forest's former natural abundance. "We are not poor—merely simple," the man exclaimed, when I asked him about the kinds of development benefits he might expect to find in his new home (figure 5.1). He quizzed me about problems with the swidden project that the hydropower project was undertaking, which were widely known, and which would be part of their lives once they moved. Even with their budget, the technicians had not been able to fix the problems, he surmised. Given what must have looked to him like an unlimited budget, it was a devastating criticism.

Elsewhere effacement took the form of an explicit naturalization as a way of evading development. I have already cited the case in which an elderly

Figure 5.1. "Not poor, merely simple." Photo by author.

Kaleng woman claimed that they were normal poor because they "live naturally." *Thammasat* (nature) means literally "in the condition of dharma," with the suffix, *-sat*, having similar semantic range of the Latin *nacio* (e.g., nation; cf. birth-right), including connotations of karmic rank; it is also the literal translation of "nation." It evokes something like "lawful condition of existence" if taken with a Buddhist sense of impermanence and karmic transcendence of the human form; dharma is not *nomos*, but it operates across a similar conceptual topology. In another lucid analysis, a woman argued that they should "live naturally, not according to science" (*tam thammasat bo witthayasat*). Exceeding her direct meaning, the phrase is an instructive rejection of expertise. *Witthayasat* means science, book learning, or secular expertise. *Witthaya* also connotes the trickery associated with the occult, the use of potent medicines, and even the dangerous but potent familiarity with forests and their non-Buddhist spirits—that is, explicitly *not* a state of dharma. Taken in modernist terms, science (*witthayasat*) and nature (*thammasat*) divide against each other, but taken in traditionalist terms, "science" is aligned with the mystical potency of animistic spirits, and together they contrast against a natural order of law, dharma, and the fate of beings that are inherently unequal in their state of grace (*boun* [merit]).

I do not think it is accidental that women have voiced these reservations. If effacement, evasion, and naturalization form an undertone of political speech that guards against state projects and technical expertise, the overtone by contrast is that of natural potency, an explicitly masculine political value that invests uneasily, but productively, in the bureaucratic state and natural abundance. Somvang, the model farmer I introduced at the beginning of the chapter, became embarrassed when we came upon a virility fetish he had erected at the opening to his rice fields. It was made of a carved wooden phallus and woven bamboo mesh to catch spirits; a lemongrass plant at its base evoked the female principle of fertility. It belied his claims to modernity—or so his embarrassment seemed to say—but testified to the complex interplay of virility, potency, and natural potential that undergird Tai metaphysics of power (Wolters 1999; Reynolds 2005).

Similarly, the abundance of nature taps into powerful cultural imagery. Khammai, a biofertilizer technician for the company, talked about taking fishing trips with friends from the company. By canoe they travelled upstream, above the powerhouse, with rods and reel and modest outdoor equipment. They camped overnight, roasting their catch on bamboo spits over a smoky flame and drinking rice wine from cups cut from bamboo. He was not alone in this enjoyment of natural plenitude—high-ranking company staffers also went on such trips. He became poetic commenting on a scene of village children who were dragging long skein nets, fishing in a muddy floodplain of waist-deep water. "Children are playing a game with equipment at which they're very skilled. But it's a game truly to pursue life, named catch shrimp, catch minnow."

What I translate here as "pursue" is an enigmatic construction, *haa liang*, a compound verb conjoining *haa* (to search out, e.g., to gather vegetables or to fish) and *liang* (to feed or raise)—"to seek-raise life." As Deborah Tooker (1996, 342) argues, *liang* is a term of potency in which the "soul substance" of masculine power contrasts with the domestic, frequently feminine element of raising and caring for living beings. The pathos here is that of underdevelopment and the "potential for prosperity" (Singh 2011, 227; Thongchai 2000).

Singh further argues that "potency" (*amnat*, i.e., power) extends specifically to "natural potential" in the context of contemporary Lao resource regimes.

While sustainability does not translate straightforwardly, natural abundance directly evokes a deep-seated cultural reference to "prosperity" (*chaloen*) that far outstrips the limited language of state-orchestrated development and in certain respects echoes the semantics of sustainability. Prosperity is a harmonizing image of collective maximization and does not necessarily reduce to programs of modernizing, progressive development. Heavy stalks of grain, lively fisheries, and rich fruit trees are the very image of abundance afforded by nature, *thammasat*. There is also an important aesthetic dimension. For example, certain villages in the area had planted beautiful flower gardens along the central axis of the village. *Chaloen* also has a strong spatial sense associated with the village as a spatial unit as well as with the *meuang* (the polity or the district), and the household; thus it carries the weight of legitimate political dominion. On the entryway to their new home, for a house built uphill and away from the flood zone, one family had written elegantly in the concrete, "bring this house to prosperity."

While often translated into English as "development," *chaloen* has little sense of temporal process and grammatically contrasts with *phatthana* (development) used for international development or economic development narrowly construed.[2] *Phatthana* refers to a process of growth and emergence, such as an organism coming into maturity or the teleological orientation of Rostow's ([1960] 1990) stages of economic growth. It took on its current political meaning under the American influence of the 1960s and was called on to translate an essentially novel political program of anticommunist developmentalism. In contrast, *chaloen* describes a state of being and does not make recourse to expert knowledge for a theory of secular social transformation. It entered the lexicon from classical Khmer, which was an important political influence for precolonial mainland Southeast Asian polities, and it implies the legitimacy of authority insofar as a ruler's dominion depends on its prosperity. Indeed, many businesses and temples include *chaloen* in their names, which is almost never the case of the term *phatthana*. Finally, while *chaloen* clearly has a long history of cultural resonance, it also plays an important role in legitimating Laos's contemporary neoliberal postsocialism, and it is a mistake to think of it as an apolitical cultural value. The affirmative term *chaloen* (prosperity) organizes complex sociocultural judgment and affectively rich voicing of how life should be lived.

The political form of ad hoc, overlapping, and heterogeneous regimes enables diverse normative agendas, which in turn are integral to dynamic projects of rule. Sustainability and *chaloen* are two distinct axes of normative value that point toward differential contexts and intricate relations of governing, but in turn they effect a hinge between transnational and Laos-specific spatial arrangements. Singh quotes the former prime minister on the question of watershed protection, in a comment easily fits within both concepts: "If we don't do it, in the future we will have concrete dams with no water. . . . If we don't protect our forests our objective of becoming Asia's battery might not come true" (Bouasone Bouphavanh quoted in Singh 2011, 228). Even though they emerge from radically different histories and ultimately have different scope and practical implications, sustainability and *chaloen* are both terms that temper "development" narrowly construed and help form a discursive repertoire in which intersecting transnational and regional regimes can be negotiated in practice. These links among cultural affirmations of prosperity, sustainability, and development were an important part of the background for the company's program of intensification.

Intensification

Risk management was the most immediate response to IRN's advocacy campaign, but it is also true that their activism resulted in a radical intensification of intervention in riparian lives.[3] The company's integrated livelihood interventions, with their own specific autonomy and generative capacities, would have never occurred without the flexible assemblage entangling the activist group to THPC. The point here is not to lay blame on IRN, which was not involved in making management decisions or conducting environmental activities, but to describe the conditions in which a sustainability enclave extended over a precisely defined geographical expanse in central Laos. Clearly, it is extremely difficult to advocate for environmental causes without creating real risks, and there may be no such thing as a risk-free politics. And yet in these circumstances, ultimately, villagers bear all the risk. Reciprocally, sustainability work introduced an often creative modelling of its own imagination of environment through a management device called a logical framework. The logical framework became the central point of an intensified anthropogenic dynamic.

The logical framework is a management matrix for organizing diverse activities in such a way that they ostensibly add up to a coherent result for environmental improvement. As it is affectionately called, the Logframe has become a centerpiece of postplanning development project work. Developed by Peter Drucker under the rubric of a technique he called "management by objectives," it was later modified into "management by results." The difference is that management by results incorporates quantifiable indicators to tell the manager whether the objectives have been met. The Logframe is essential to a culture of outputs; outputs are by definition quantifiable indicators of achievement that govern the work involved. I take the Logframe as a critical artifact of the vernacular materialism or ontopolitics that has helped define sustainability work writ broadly, including that of Lao hydropower development. The Logframe is a material-semiotic form designed for navigating uncertain worlds and, from within the logic of business rationality, a key demonstration of a generative relation to threat and opportunity that functions performatively, not unlike Richard's embodied practice.

One can take classes or watch instructional videos on how to develop a Logframe, as I have done as part of this research. I also fielded periodic requests for insight into how Logframes are supposed to be interpreted by different informants who thought I must have some privileged relation to this bit of management mana. Of particular importance is the hierarchy of "levels" which give sense to the nested objectives of the program and attempt to define the overarching purpose of the work at hand. "The project has one *purpose*," reads one training document retrieved from an archive in Vientiane.[4] "The *purpose* is not a reformulation of the *outputs*." Here, the italics (their emphasis) identify technical terms pertaining to the matrix itself, especially concerning the hierarchy of teloi that extend from activities at the most basic level, through project outputs, purpose, and goal. The "logic" of the logical framework designates a kind of algorithmic process of following instructions governed by future objectives. Following rules (instructions) could be viewed as a wager governed by a possible achievement, not unlike using a recipe to cook a meal.

As a postplanning device, the organizational form is supposed to accommodate failure and learning. Given the difficulty of knowing precisely how to act, it is better to get to work with a "good enough" understanding of the situation, make some mistakes, and revise as needed. In this case "postplanning" is configured as a technique of working with open-ended contingencies. As Halpern writes in her discussion of the seminal cybernetics paper "Behavior, Purpose, and Teleology," encoded purposiveness involved a critique of deterministic causality: "in this new 'purposeful' and behaviorist approach there is a cause and effect, but the relationships between cause and effect are not predetermined. Rather, they are a choice, a likelihood, the result of a series of possible interactions" (2014, 46).[5] It is a useful technique for managing uncertainty. In the context discussed here, the Logframe was a tool for stimulating personal motivation.

When it came to getting busy in the villages, I was primed to expect considerable energetic work on the part of the company's environmental technicians, several of whom I first met during the external performance evaluation discussed in the previous chapter. What I did not expect was the degree to which the management device framed their day-to-day work, especially with reference to quantitative targets. The Logframe set up a reiterative process in which the technicians needed only to keep their eyes on the numbers of activities to be achieved. I had Bouali, the program manager, explain to me how the Logframe worked even when things go wrong. "In that case, it's ok," he said. "Because we know what we need to do according to the schedule. Not everything will work perfectly. If we don't finish one month we can move it to the next month. We don't have to spend the allotted money by a certain date. That way, we can go more slowly or quickly as we need to." If the numbers usually were reached successfully, it demonstrated the recursive practice of self-monitoring in which achievements became an explicit source of self-worth. But it also meant the targets were defined narrowly, with the objective to finish many instances of an activity and then later gauge how frequently it was successful. This coupling of numerical performance, managerial flexibility, and cultural repertoires of effectiveness was aptly demonstrated with Suchada, deputy manager and energetic field hand.

At fifty-four years old, Suchada was something of a paragon of motivation and commitment. He ran on caffeine and cigarettes. His face and arms were browned dark, so you believed him when he said most of his time was spent working with villagers in the field. He wore shorts, a rarity for a man of any authority in Laos. A monk for thirteen years in Vientiane, he left the monastery after the Americans withdrew in 1973. One morning when I was working with him, the night before he and another technician had raced off to a project village by motorbike to fix a broken irrigation pump before the heat of the day could kill off a season's vegetables—returning only around 3:00 a.m. We spoke Lao interlaced with English; Suchada had forgotten the

French of his youth to learn in his forties this new language of currency. He spoke quickly, but not rushed; his voice was sonorous and clear, yet his movements were caffeinated.

Suchada described how the Logframe allowed him to gauge visually what he had already achieved and what work remained to be done. The tasks were discrete, measurable things, things that could be counted, achievements that required no comments to elaborate. Even if others were not so practiced, they too were intently focused on achieving project outputs. The motivation of the Logframe simply enforced the rapid cycle of activity-reward-repetition, and I wondered whether measuring dopamine levels could be an interesting way to expand my methodology: Numbers of water pumps installed in village gardens. Hectares of newly irrigated land. Households incorporated into the rice intensification program. Numbers of families current on their deposits into the village microcredit scheme. Scripted within these metrics, but hardly determined by them, Suchada's precision and skill were manifest. When he talked to villagers, one could see that they trusted him, his personality mixed with a narrative vision that conveyed the program's promises of improvement. His movement frequently displayed a deft elegance, as when he once snatched a rabbit up off the ground, flipped it on its back and precisely spread its genitals to determine its sex. Whatever the Logframe's powers, it could not but hold a vast debt to these kinds of familiar and charismatic capacities. Without them it was an empty shell.

Such an abstract system, arising from American management practice against a background of behavior modification techniques, information theory, and algorithmic logic, raises a wide range of potential questions when interpreted in the context of Laotians' sustainability experiments, only some of which can be addressed here. First, it is clear that this approach was essential to the motivation of company environmental technicians, but motivation is construed only as objective behavior and does not imply questions of interiority or subjectivity. Furthermore, the *longue durée* cultural logic of shifting Lao interpretations of nature is embedded into the logic of practice, not overridden by Euro-American understandings. The Logframe provides form to that locally specific content. Lastly, the Logframe, no doubt *because of* its behaviorist logic, is quite different from the immutable mobiles described by Latour (1987). Immutable mobiles are constructed objects (including facts) that work hard to maintain their own integrity or durability as they travel in diverse contexts. Rather, the Logframe is a "flexible body," to evoke Emily Martin's 1994 phrase for a certain kind of managerial openness that emphasizes adaptability, revision, and the incorporation of difference. It is a technique for figuring out how to operate in vastly diverse contexts. I want to explore this openness, tied to motivation, through an analogy with Marilyn Strathern's (2000, 2006) understanding of audit culture.[6] For the motivated subject is an essential figure for neoliberalism's adoption of expansive, postnatural ecologies.

Bubbles of Motivation

The transformation from plan to postplanning device, as I have argued, involved a wholesale change in the role of empirical knowledge. This transformation was captured by John's lengthy narrative about translating the company's environment plan into the logical framework:

> When I arrived they had a big, ten year plan. It was good, but it didn't hold together well logically. It shouldn't take a minor in philosophy to figure out what we're doing—that's what happens when a plan is put together by scientists without any management expertise. I said, 'Can we get this onto half a page instead of two hundred pages?' We needed to show the issues and initiatives for each [activity]—not change the detail, but changes in the methodologies, delete some activities and add others. Now it's very different from when we began. We weren't fully aware of what we were trying to do, and the beneficiaries' perspectives have changed too.

In fact, the original plan, which was fifty pages long, had scheduled to roll out predetermined activities across a ten-year work plan, with some margin of error for costs but no way to incorporate what would inevitably be a dynamic interplay of possibilities, experience, and failure. The Logframe by contrast was seven pages long; trial and error was built into its very form.

By contrast with the original plan, the Logframe made virtually no claims about the empirical situation at hand. To be sure, it is possible to read between the lines, but the point here is not merely the loss of complexity (Jensen and Winthereik 2013, 114). As Strathern (2006) shows, to argue for a loss of complexity is a banal critique when practices such as audit are explicitly concerned with reducing complexity. The ten-year plan may be reductive in the way it characterizes its object, but the Logframe does something else altogether—it describes how the company thinks of itself and its activity. Strathern writes that "auditing practices make sense of an organization, then, by requiring that it 'perform' being an organization" (2006, 192). Analogously, the Logframe makes sense of sustainability not with reference to the authority of positive empiricism, but through the teloi of its performed achievements. It says, merely, "this is what we are doing." A collapse in the presumption of naturalism is involved in this shift toward operational self-reference. The Logframe is not reductive; it is purposively self-referential and minimalist.

Over that year of my fieldwork it became clear how much work was involved in rolling out these diverse activities in village after village. Undoubtedly it was straightforward enough to enlist a handful of farmers in one village into trialing a gardening technique or orchestrating an investment in raising pigs. But in each case, once underway the process had to be backed up with ongoing technical support and troubleshooting, and, as Suchada discovered, new troubles would emerge each season just as they were expanding into yet more villages. Motivation and the threat of failure confirmed each other. For his part, John had a remarkable habit of bragging about the effectiveness of his approach, going so far as to name the management tactic after himself. "The Harriman approach—carry a big stick and a fistful of cash, and after they do it the way I told 'em, they realize it's the right way. That's what I call participatory development!" In another case, he bragged that it had been called the best-managed project in all of Southeast Asia. One could write off such braggadocio, except it seemed consonant with the entrepreneurial ethos, not to mention the performative masculinity, in which constant encouragement and affirmation was more for the sake of himself and the environmental technicians than for any outside observers.

From whence comes this concern for the feelings of environmental labor, which, after all, was a primary concern of the environmental consultants discussed in chapter 4? On the one hand, the Logframe is explicitly meant to provide a recursive approach toward failure. As Rob van den Berg (2004) argues, the Logframe emerged as part of a culture of organizational learning in which management could think of itself as adaptive to the most improbable conditions. Learning means being o.k. with getting things wrong—just try harder next time! Rational planning contains within it the seed of its own demise, namely the impossibility of complete empirical reference needed to ground the validity of its practice. It thereby performs authority (law, rule, transparent implementation) as a supplement. By contrast, this achievement-oriented managerialism hinges on a different impossibility, namely that with enough initiative or motivation the right recipe can be hit upon. Managerialism is plagued by the threat of failed achievement; it performs gumption as its necessary supplement. These are, in turn, two different ways of viewing rules—one in terms of authority and obedience, the other in terms of instructions (algorithms, recipes) and manipulation.[7] Following the rules can be opportunistic.

Is not bragging just another way of calling attention to certain achievements rather than others? Recall in the first pages of the book I relayed a passage of dialogue from Richard in which the trending of indicators was meant to take the place of rigorous data collection and analysis:

> *Consultant*: Ok, so you just have those sets of data—from what I've been told there's no analysis that's been done yet. Because you don't have enough information.
>
> *Richard*: We're—we're—we're—we're *trending* it to see if there's anything. And on some indicators it looks like yes, [unintelligible]. On malnutrition under five, that indicator seems to have been a huge improvement in villages we've started work programs in.
>
> *Consultant*: Ok that sounds great![8]

Far from being the sort of straightforward learning device that it was meant to be, the Logframe functioned as a quantified stimulus that dispensed small tokens as a reward for energetic activity. "We want numbers, not comments," John told me, speaking of the monthly reporting requirements that Richard had put in place, which were explicitly minimal. "The Logframe provides a standard form for enumerating activities. A lot of what the staff achieve, there's no need for comments. You have to get the numbers down, focus on the numbers." Later, as the program suffered a series of setbacks, John admitted the focus on numbers had been counterproductive. The field technicians were burnt out, and many of the activities simply were not working well. "It's not the management methodology of using the Logframe, but perhaps the way I may have led the staff to *believe* performance targets are important," he emphasized. "When we have a meeting we just tick everything off as we deal with it. It's true we were rushed, we had activities underway in 45 villages.... They're smart people, motivated. That's how we've managed the Lao staff, how we've created ownership." This kind of quantified, algorithmic intensity created bubbles of motivation taken up by environmental technicians and managers.

And who exactly is doing the manipulation? Strathern discusses the "awe of technique" through which organizations produce amazement at their own self-descriptions: "The new accountabilities use a false mirror device. For they require that on the shield be painted not just a depiction of ourselves produced in order to impress others but *a picture that shows how impressed we are with ourselves*" (2006, 188, her emphasis). Too much criticism, or criticism of the wrong kind, and motivation collapses; in this sense, management works to preserve autonomous spheres of intensification, little bubbles of motivation inflated by the constant performance of success, narrowly construed.

Notwithstanding such a pyramid scheme of aspirational achievement, the environmental work created practices that took on their own forms, independent from the focus on external performance in the service of risk communication. How could it be otherwise? Technicians were tasked with rolling out sometimes fanciful project activities that implied serious changes to villagers' lives and livelihoods. Those activities were framed by the intractability of deep-seated problems poorly addressed by their work. Compounding the challenge of complex and difficult experiments in livelihood changes, the river's ongoing transformation constantly introduced new elements and cut off existing possibilities. For villagers, the impacts of the dam were frequently a dull roar of slow and subtle uncertainty and progressive impoverishment obscured by daily events and other powerful regional trends. The original effects of the dam, now familiar, resonated discordantly with the increasingly intimate attempts at livelihood improvement. The crush of aspiration, villagers' respect for the environmental technicians, and the onset of disappointment were particularly acute as all three forms of rice intensification systematically failed over the course of the year. In the remainder of the chapter I discuss these in turn to explore core processes of ecology, economy, and state through the temporalization of this anthropogenic river.

Ecological Wasting Processes

Hydropower organizes water within a field of gravitational difference. In doing so, it reconfigures the temporal interplay of soil and water. "The scheme's whole justification depends on switching almost all the dry season flows to another (low elevation) river," one consultant wrote, advocating that the company should "manage the two new river systems so that they develop productive and diverse new ecologies, with ecosystems similar to those of wild rivers with the same flow patterns."[9] The novel riparian ecology was a function of completely altered riverbed topography, fluctuating seasonal and diurnal hydrological flow, and reduced sunlight conversion by aquatic plants due to the dissolved solids. Along the lower elevation river, where most of the affected people live and the most severe ecological impacts have occurred, the banks have continued to erode since plant operations began in 1998. The steady wasting of the riverbanks displayed a temporality of irreversible transformation.

Unexpectedly, an important component of the erosion takes place during the dry season. Douglas, the geomorphologist whose work I described in chapter 4, showed that the erosion is due significantly to the pattern of electricity consumption in Northeast Thailand. Across the border, during the dry season, the dam supplies energy during peak demand, when electricity use is at its highest. In the mornings and evenings, the power plant goes into full operation; the water discharged from the power plant fluctuates twice a day from nearly nothing to full flow. A regulating pond dampens some of the extremes of this punctuated diurnal pulse. However, tied to the daily pulse of urban electricity use, the pulse of water discharged into the river rises and falls at the same frequency and amplitude. The drying soil "sloughs off," Douglas explains, into the swirling waters. As the exposed, wet layer of soil dries out, it crumbles into the water. This repetitive wetting and drying, not merely the mechanical action of the water's current, characterizes the unstable relation between soil and water.

Imagine the secret connection between the habits of electricity across the border and the diurnal sloughing in Laos.[10] Rising for work, then returning home again in the evenings, Thai consumers make demands on the power grid, which accumulates demand into a mass index of Thailand's industrialized agrarian capitalism. Ebbs and peaks in the demand for energy provoke a covalent oscillation in power generation, and the power station releases a flood of water as it works at peak capacity. Wetting and then drying, the soil crumbles in an inadvertent and irreversible wasting process.

Before the dam was built, the oscillating flow was seasonal; now the diurnal pattern predominates. The ecological patterns have shifted. Overall the river had become much wider and shallower, especially further up the watershed where erosion was at its worst. Farther downstream, perhaps a quarter-million tons of sediment pulsed with each flood into new configurations of riverbed topography. In addition to the impacts on paddy, the riparian fish ecology had become totally transformed. Rapids spread thin and sandbars migrated season on season into unstable configurations. Deep pools that once were essential fishing grounds were filled with sediment, and the water itself carried a high sediment load, sharply cutting down photosynthesis at the base of the aquatic food chain. Tasty bivalves and snails disappeared, smothered by the waves of sand; new species of fish have taken over while others—the list of local names is long—appear now only sporadically. Eroded riverbanks are scoured of plants; in other areas, unfamiliar water plants choke the now-shallow trenches of the riverbed.

In Nha Sai village, the imposition of the dam shut down a vibrant group of professional fishers, about seven people, plus traders who carried the fish downstream to the road and to Thailand for sale. The change was sudden—too much water and silt. Before five deep pools and rapids afforded fish habitat and tactical opportunity for certain nets and traps. Specific fishing techniques, which require certain topographical features, no longer functioned. A refrain repeated in many villages was that the water flow is too strong. Too much water now. Unable to run nets across the river. The flood is muddy. Water runs too fast and there are no deep pools.

In certain villages, large tracts of important crop land were increasingly untenable for wet season rice. Villagers referred to certain plots as "risky paddy" (*naa siang*) (figure 5.2), a neologism that echoed more conventional kinds of cropping like "irrigated dry season paddy" (*naa saeng*) or "normal wet season rice" (*naa phi*) but with a temporality that pointed directly toward the challenge of making difficult decisions about where and whether to plant.

Farmers argued that flooding, which is an integral dimension of the paddy ecology, damaged the crops in two distinct ways. First, it meant that certain tracts of land experienced longer floods that failed to drain quickly enough. Normally when the floods retreat, farmers use paddy bunds and channels to control the timing of that drainage to provide the right balance of nutrient

Figure 5.2. Risky paddy, Nha Sai village. Photo by author.

supply and moisture. Now the flood lasts eight, ten, or even fifteen days, whereas before, they reported, a regular flood would last four to six days. Second, the suspended sediment particles significantly affected the viability of the rice plants. Clean water can flood the fields for a lot longer than muddied waters before the plants die. The silt actually clings to the stalk of the plants, inducing a slimy bacterial growth. Increasingly, these fields were being abandoned. In one village they had not planted certain fields for the previous three years, and the floods had been moderate. The next year they decided to plant, only to face particularly extreme inundation.

The viability of rice fields, rice plants, and specific strategies for cropping are directly implicated in the changing topography. The term "risky paddy" came to be used a few years after the dam. In 2001, Nha Sai village lost 70 percent of the rice fields to flooding, making clear the changed temporal interplay of water and soil. Previously, they expected every two or three years a flood lasting longer than fifteen days. Fifteen days was the cutoff: about half the rice would drown at by that point—the water was clear and many plants could survive. But in 2001, the water was muddy and silt coated the rice, killing most of it. The next year they abandoned twenty-five hectares, about a quarter of their land, because it was especially flood-prone. The remaining seventy-seven hectares—the risky paddy—they plant every year regardless. In 2002, the harvest was lost. The next year, there was little flooding; in 2004 everything was lost.

So much living happens in this unbounded world where the land-water distinction fails to rest, this patterned riparian topology of water and soil. The floodplains here are fairly flat, so one could imagine a map of the area with fields closer to the river progressively more likely to flood too long for too often. In fact, an NGO working independently tried to make such a topographical map as part of a livelihood security program, but found that the elevation changes were too subtle for the GPS sensitivity they were able to work with—for the water, ten centimeters can make all the difference, but that difference was too tricky to measure using the technologies at their disposal. As I worked through these villages interviewing people and myself negotiating this transient aquacline, the landscape itself created a real life instantiation of the emergent risk landscape. Every planting season those farmers must make decisions about which fields might or might not flood that year, calling attention to agricultural practices that are not obvious, honed to precise environmental possibilities requiring years of observation and experimentation. And yet the river will continue to erode for decades, foregrounding the unpredictability and impermanence of anthropogenic natures. Like the covalent oscillation linking Thai electricity demand to Lao erosion and sedimentation, the production of uncertainty oscillates between multiple fields of relation—from villagers' relation with their paddy, to the overlapping moral and political authorities of the sustainability enclave, to the intractable problems whose dimensions remain unstudied and conjectural.

Uncertainty is not the opposite of "certainty," that elusive telos of pure reason. Nor is it a thing, a quantity or even a state of being. Uncertainty is a relationship that indicates the value of knowledge. Risky paddy marks an affective register in which hard-won knowledge of local specifics of field and riparian flow is called into question by the literal undermining of a riverbank. The work of figuring out a new riparian topography is exploratory and subjected, in which novel obligations to environment take shape and make demands on established practices and ideas. Even if general biophysical impacts are roughly apparent, the specifics call out an open-ended future in which people are forced to imagine the different ways dangerous or potentially lucrative situations may play out. In this context, it makes no sense to align "risk" or "uncertainty" solely within an epistemic genealogy or discourse. Clearly these terms have an explicit semantics, and this can and must be traced historically. Yet it is also true that reality becomes deformed by the event, and the anthropogenic river exceeds discourse, sometimes forcing thought onto new objects.

In the meantime, for villagers, the concatenation of problems dominates. The details matter, continuously, and working with things constantly threatens achievements that do not come to pass, ambiguities that are never resolved or relations about which little is known. Even years after the dam was built, rice seedlings are buried in mud with unpredictable floods. Fishing possibilities have collapsed. The water flow is too strong to use their conventional gill nets. The steady sloughing of soils from the riverbanks induces cascading transformations that, for all practical purposes, are irreversible and unpredictable. The exploration of new possibilities—cramped, opportunistic, and limited—proceeds apace. Migrant labor to Thailand became a constant feature of family life. Young men migrated seasonally for agricultural work, while young women went mostly for household labor or trade (in at least one case, prostitution); both are subject to exploitation, physical violence, and wage theft. The tenuous viability of wet season rice ecology along segments of the river also created new demands for experimenting with higher risk and potentially higher reward modes of intensified production, especially during the dry season when flooding was not a risk. While the destructive opportunism of large dams undermines the intelligibility of other worlds, the promises of entrepreneurial sustainability and market-oriented redemption opened onto what Lauren Berlant (2011) calls cruel optimism.

Cruel Optimism's Economy

Exacerbated flooding and the loss of wet season paddy played directly into the company's attempts to greatly expand rice production, even while the company denied any direct responsibility for the floods. Along the central stretches of the Hinboun River, farmers invested considerable resources, at the company's behest, in green revolution-style paddy intensification strategies such as dry season irrigated rice and, in the wet season, improved varieties of rice with chemical fertilizers. Wet season rice cropping took to comparatively moderate supplemental technologies, such as adding fertilizer at key moments of the crop cycle or using enhanced seed varietals. Yet the ongoing transformation of the river placed tremendous emphasis on their experiments with dry season rice, for it was the possibility of benefitting from a second harvest each year, without the threat of flooding, that captured the imagination of farmers and technicians alike.

"Optimism is ambitious," writes Berlant (2011, 2). Not all optimism is cruel; rather, cruel optimism is that in which the object of desire actively impedes one's flourishing. Cruel optimism is an artifact of managerial entrepreneurship and the temporal conjunction of promises, hype, and material potential, what Berlant describes as a "melodramatic view of individual agency [that casts] the human as most fully itself when assuming the spectacular posture of performative action" (96).

Hinkham village was paradigmatic of the way in which the stuff of sustainability, and the collapse of optimism, played out the temporalization of anthropogenic rivers. Previously they only farmed in the wet season, with a good yield in the range of two or more tons per hectare. The year before I had arrived, several villages had impressive five-ton yields using TDK, a popular seed brand, fertilized with a mix of urea and ammonium nitrate, all imported from Thailand, subsidized by the company, and distributed as a loan to be repaid into the village fund. This year, a big push was in play to expand intensification to more villages. John and Buali targeted a seven-ton yield; others were incredulous. One farmer joked about a Professor Saisomboun (the name is very auspicious) who had once gotten an eleven-ton yield. I asked further; the professor turned out to be imaginary.

Keovilay, a young farmer who had already been doing some dry season rice, had enlisted the previous year in the dry season rice project as part of a pilot in his village; the year of my fieldwork it was expanded to most of the village households—about forty. This entailed fertilizer, seed, and subsidized electricity for irrigation; the cost of the fertilizer and seed would have to be repaid into the new village credit system. He had already been doing some dry season rice, which was part of the reason he had been enrolled earlier than other villagers. Clearly he was eager at the outset, speculating aloud about being able to buy a hand tractor if the crop was good. Later, when the company technician made his rounds, they argued: the seed given to them smelled bad out of the bag, and it never sprouted. The technician suggested he had let it soak too long or bruised the seed by treating it roughly, but Keovilay and his neighbors rejected that argument and demanded new seed, only some of which germinated. The company insisted on a particular high-yield variety, TDK 4, whereas before they planted two different varieties as an easy strategy of diversification. The fertilizer had also been damaged—the fifty-kilogram bags of powder had turned to hard bricks, which he and his neighbors crushed using hammers and applied as best they could, the fertilizer still in chunks. Then the rice grew beautifully with lush, full tillers, but the plants did not flower and therefore proffered very little grain. His fields yielded one ton per hectare, a far cry from the grandiose promises that had been made.

Many farmers like Keovilay readily embraced the possibility of commercial production through enhanced methods. Their mode of affirmation linked commitment to expertise and investment with an elaborated relationship to the company and its technicians built on promises of efficacious expertise. By the same token, describing these relations in terms of threat and opportunity leaves open the ambiguity among hype, opportunity, and achievement and makes clear that commitment to a potential future is also a commitment the power relations through which it may or may not come to pass. Empirically, people routinely affirm power relations, and they do so *in time*, that is, not in the abstract or in terms of a pure judgment, but in terms of the unpredictability and irreversibility of relationships that are difficult to judge, that may turn out otherwise, and in which relations among people and things are continuously wrapped up in each other. These kinds of techniques just might work, after all.

Keovilay was convinced the fertilizer was the problem—they had been instructed to put on too much, and they were not able to spread it evenly. That is how imagination and uncertainty work here, when you are trying to finesse a cruddy product, smashing these bricks of chemical fertilizer, which should be a fine powder, wondering if it will dissolve and melt in the soil or if it will burn the plants when you apply it.

But the failure of the plants to seed had happened across nearly all project villages. The company line, which the technician was adamant about, was that it was a regional problem—some parasite or infection had taken out this variety of rice across northeastern Thailand and central Laos. Keovilay laughed heartily, or maybe it was bitterly, when I suggested perhaps next time the advice will be better—"He won't be back," he said of the technician. "He said we might get 10 tons!" Yet the debate with the company had shifted. Saiphong, a field technician charged with ensuring villagers make payments into their village funds, had come by several times, badgering them to repay loans for the inputs for the failed crops. The villagers mocked her in gendered tones. "She's like a busybody auntie trying to find out what you spend your money on," one said. They simply did not feel responsible for paying back the loans, but her argument was different: "if they don't make payments into the fund, where will they borrow money next season?" A former village chief, who had been involved in setting up the project, showed me a list of all the farmers and their debt commitments he had signed when the original disbursement were made. He expressed regret for helping sell the project to his village. "We believed the company and followed all the directions—now there is no trust."

Similar experiences predominated, even in villages that had been growing some dry season rice before the company got involved. In Nha Sai village that year, the beautiful plants grew chest high, or even higher, but then died back just before flowering. Farmers wondered about a red worm that had been found nested at the joint of many plants' tillers. Technicians from the company told of a field trip they had taken, with sixteen village heads in another province, where similar die-offs had happened, offered as evidence that it was not the company's fault. The explanations were met with incredulity, irony—and serious attempts to explore alternative livelihoods. "If I can say this openly, we didn't believe the company technician," said one man. Another: "No one here agrees [with the company's assessment]. But everyone here says they agree." A number of youth had left for Thailand or Vientiane for work; other families had invested in tobacco farming under contract. There was also palpable worry about a new eucalyptus plantation concession that was cutting into village territory, in effect squeezing them between appropriation of forested hill lands and the continuously eroding and flooding riparian zone.

The anthropogenic river thrusts people's optimism into new domains of negotiated technical and economic expertise, in which the value of that knowledge is constantly at stake. Agrarian failures like the dry rice experiment are frequent features of the anthropology of development, which has noted that development is the "management of a promise" (Pieterse 2000) and even the "management of the third world" (Escobar 1988). Sustainability is no different, and this combination of anticipation, instrument, and predicament makes sustainability a way of life, even for comparatively affluent and well-situated farmers like many of those in Keovilay's village.

The anthropogenic, as it were, always occupies that compromised position between ecology and economy, and therefore it is not much of a critique to say that the promises of sustainability ring false, for one does not get out of it that easily. People's lives take form in its interstices, and its managerial forms are an integral component of the ecological contemporary. Likewise, failure is more interesting than much academic research has let on, for it demonstrates not only the agency of things but also how material relations take form via expectation, obligation, anticipation, and commitment. Working with things takes time; requiring skill, it is often unfinished and demands commitment to the material specificity of one's predicament. Failure is the norm. The stuff of sustainability—cruddy fertilizer, mouldy seed, a debt schedule, arguments with experts—resonates against the equivocation of trust and expectation. The cruel optimism of entrepreneurial life consists in this ready experimentation on new techniques, involving intimate and formative forays into new obligations in which the stuff of sustainability is wrapped up in the hype/promise of achievement.

Effacement and the State

"Effacement"—what I name the evasion of complex power relations while creating cramped spaces of getting by—shows that harm is not merely a physical effect or an impact of late industrial projects. To become a harmed body means becoming a person capable of bearing harm, which can be understood as a capacity powerful in its own right.

Along the headpond, rice intensification took its most extreme form: the attempt to convert hill rice shifting cultivation to permanent rice cropping, whether through "permanent swidden" (*hai ḳong ti*) or what villagers called "paddy-style" farming (*paep na*). Both terms are ironic neologisms; like risky paddy, they mark odd entities—new things in the world, nonviable experiments, intractable situations. Permanent swidden is an oxymoron; it refers to shifting cultivation that has been fixed in place. Fields previously used once every six or seven years must be used repeatedly. Paddy-style fields are hill rice fields that have been dug up to create low bunds or "terracing" meant to catch water, but are nonfunctional in practice. Thin, sandy soils and the general absence of flowing water suitable for irrigation had long enabled a strategy for living that emphasized sparse settlement, long rotation cycles for hill rice, and an important role for extensive riverbank gardens and vibrant fisheries. That combination began to unravel as the Lao state was brought in to shore up the failures of the sustainability experiments.

Now in the same fields for the second year, the weeds were oppressive and there were few nutrients to sustain the rice. The plants flowered but the grain was "feeble" (*orn-ae*)—empty husks or grain that crumbled into powder when pressed with a thumbnail. In order to implement this program of stabilizing shifting cultivation, the hydropower company worked closely with the district government to quickly set up the latter's land allocation program, creating a conjunction of powers that gave the lie to the ideal of private sector intervention. When it came to enforcement, the privatized regime simply relied on district officials to give the plan weight of law. "We're like the mother," John, the environmental manager explained to me. "The district is the stern father." Framed differently, he evoked more directly the neoliberal stakes of the management regime when he said the company was like the carrot and the district like the stick. For some villagers, it demonstrated that the most powerful affirmations sometimes can take the form of explicit rejection and refusal. For others, their tactics were subtle evasion and deflected political speech. Here, people explored sometimes dramatic alternatives to contend with the bitter mix of ecological, economic, and political relations.

In Somvang's village, large quantities of forest products, especially rattan and *ķisi*, a tree resin, were stacked alongside the road or in covered sheds, evidence of an energetic rush of forest gathering activity that was strange at the height of the rainy season, when farmers should have been working in their fields. One family had taken to gold mining in the sandbanks of the river, digging in dangerous pits amid the river-worn boulders, an activity widely taken to indicate their desperation. Earlier that year, in a willful, deliberate rejection of government policy, forty-two farmers had been arrested or fined for pioneering swidden in the nearby protected forest.

In adjacent villages similar problems were underway. The chief of one village, Buathong, was visibly distressed. When we first met, he asked me within minutes whether the company would consider abandoning the environmental program in favor of a simple cash pay out to villagers. The improvement activities were intensely disliked by many villagers—as in Somvang's village, the farming interventions had largely failed. Farmers described in detail the collapsing productivity, an influx of rodents, and increased reliance on forest products sold for meager cash to itinerant traders.

Slowly it emerged that a group of villagers had submitted a petition to the company to quit the company's environmental program.

Buathong himself was unable to sign the petition because of his official position, and he feared being sent to reeducation "seminars" in the district center (during the height of the communist period, one of the main reeducation camps was in this district). He had been handpicked for the position on the basis of his good behavior, approved by the villagers, and then inducted into the Lao People's Revolutionary Party—against his wishes, he told me, although no doubt he was flattered at the time. Now he spent all his time in meetings and was not even paid enough to cover travel expenses. He had no time, he kept repeating, using the phrase *nyoung yak* that combines a sense of pervasive anxiety with a condition of being harried, as if a jumble of inadequate options or obligations all come together at once. His visage brightened momentarily in our discussion. Being in the party would not be so bad, he said—if it were not for the district's policies. Whatever the case, his apprehension was visible in his collapsed posture. We talked for hours into the night, since he hoped I might have some means of interceding. He surely sided with the villagers who put forward the petition, but it put him in an untenable predicament. A good chief is one who is honest, who tells the government what is necessary, he told me, looking crushed. A good chief is resolute enough to insist on what villagers need and to withstand the hours of meetings in which officials work to undo their resolve.

When I asked what the government officials would do, he initially balked at answering. They were afraid of the district, he said. He was not sure what they would do, and he worried about his family. He spoke of the district officials using the words "potency" (*amnat*) and "control" (*bangkap*)—heavy, metaphysical terms expressing intense power. Nothing like this had happened to him before. The petition had been submitted to the hydropower company, yet the company had handed the problem over to the district, and officials were pressuring him to bring villagers into line. The conjunction of overlapping regimes in the sustainability enclave meant that the environmental program could appear as a private sector, participatory intervention from the outside, while it was largely compulsory through district government policy when it came to implementation and enforcement. By the same token, district policies took on a force they would never have achieved without the company's resources, encouragement, and motivated labor.

Questions of voice thus became key to the complex of power relations and collapsed expertise that came to define the riparian environment. Another village nearby demonstrated this in inverse form. Here the chief's house was dilapidated, made of bamboo and listing in a way that spoke of his poverty. Like Buathong, he claimed he was too busy to plant rice, and his village was experiencing "all of the same problems" with the mitigation program. Seated on the bamboo floor, where we would sleep that night, while the man's wife prepared some food for us, the interview did not last very long. For a moment the stench from cooking was unbearable, but busy with the detailed interview, I paid it no mind. The chief was adept at dispensing with questions and, as I asked about village incomes and land use, he displayed a remarkable ability at making the numbers look plausible. This kind of life, he said, is "most difficult" (*yaķ thi sout*). He had little interest in elaborating on his problems but the "local undertone" of his speech was on full display.

In contrast to Buathong's village, he insisted that his village did not want to stop the mitigation program. Everyone participates, he said; they like it—although some households such as his did not have enough labor. Many people now had to spend much of their time hunting in the bush around the village. Everyone accepts the company's program, he said; however, they had tried the program for two years and they were still waiting to see the results. Of course, he said, if they were allowed to they would definitely return to their previous swidden methods. His strategy of effacement thus centered on affirming their willingness to participate, while alluding explicitly to alternatives—such as returning to prior methods—or the technical difficulties experienced. "Difficulty" must be understood as both a technical and a moral discourse. The message seemed to be "the hydropower company is making our lives miserable, and it is your responsibility to listen carefully and gauge the precise dimensions of this impossible situation."

The chief's wife, in a gesture that spoke to our status as guests, had prepared for us what may be the classic Lao dish, a minced meat salad called *laap*. Working by hand, the cook methodically minces the meat into the smallest possible pieces—a laborious and time-consuming task that makes the forearms ache. The meat is then cooked and combined with ground, roasted rice as well as chili and other spices, making an incredibly spicy salad eaten by hand with balls of sticky rice—a sumptuous meal. Yet this particular dish, as she had made it, was both crunchy and had a curious, greasy flavor.

The meat, it turned out, was bamboo rat, and—as if the woman had recently reviewed James Scott's (1990) *Domination and the Arts of Resistance: Hidden Transcripts*—the crunchiness was the many bones that were too small to remove from the creature. The stench we had earlier experienced, I learned, was the woman roasting the hair off the rat's body.

Sometime later, when I returned to Buathong's village, it was denied to me that there ever was a petition, even though I had been there right after it had been filed during the most stressful involvement of the district. Buathong, no longer chief, had left the village to help his brother harvest. While he would ostensibly return, his house was for sale. Moreover, everyone now wanted to move from the village; three families had already left. The yields had been as low as about a hundred kilograms of rice per hectare. They asked for help from the project for moving expenses but had not yet received any answer. Many were *waiting to see* what the new program would involve—they had been promised changes—and worried that the rice-for-work schemes of self-improvement would continue to demand too much of their time. Like families described earlier, who quietly naturalized their poverty in order to avoid the potential disaster of state development, "waiting to see" indicated a temporality of averted recognition that deferred the hyperactivity of motivated participation.

Failure shrouded the project, and while I prepared for departure from my fieldwork the managers made plans to overhaul the mitigation approach. Yet the relations of domination and ecology that emerged in these villages—although not all of them—rendered living increasingly implausible. Its most potent symbol for me was not the harassment by district officials but the large, prominent billboards set up in project villages, meant to replicate the managers' logical framework planning matrices, which quantified the participation of project-compliant households (figure 5.3). If the essential meaning of an idea depends on the permutations it takes when put into practice, then the motivational logic of the Logframe cannot escape its oppressive implications. In this conjunction of multiple lines of domination, effacement is a tactic of endurance.

Effacement is only one possibility for the creation of viable conditions of existence, and it sits in contrast to the entrepreneurial practices of Keovilay or Somvang because it is concerned with pervasive threat rather than opportunity. That cramped and tenuous endurance of threat, an anthropogenic condition par excellence, is essentially creative, and much of that creativity

Figure 5.3. Village Logframe billboard. On the left is a list of management activities directly tied to the Logframe; most of the numbers are the quantification of project-compliant households. Whereas technical staff were definitely motivated by the numbers, most villagers emphatically were not. Photo by author.

rests in vital knowledge of evasion and deferral. The temporality of waiting, the muted or deferential criticism, the juxtaposition of harm with tacit assent to domination, and the subtle attempt to negotiate with the hydropower company while avoiding the state are only too brittle and inadequate. Yet they formulate a real capacity for bearing harm. They also cannot be separated from the migration, bush hunting for small game, or windfall collection of forest products that represent the stark face of dispossession and precarity. Together these form a concrete rationality for appropriating and affirming thin possibilities for maintaining life along anthropogenic rivers.

Conclusion

This chapter has combined an analysis of environmental management practices with a description of pervasive uncertainties from the vantage of Lao farmers and fishers whose riparian lives continue to be transformed. By linking techniques of managing uncertainty with the widespread contingencies of village life, I have attempted to underscore how a power plant drainage has made the rural ecology into a late industrial environment. Furthermore, I have argued that uncertainty is a feature of the late industrial environment itself and cannot be reduced to a regime of discursive power. Just as built environments and genetically modified organisms can be understood as artifacts of knowledge, so too the inadequacies and failures of knowledge can be understood as part of the riparian ecology itself.

The kinds of ecological effects experienced in the wake of the dam project are both subtle and pervasive. While large infrastructure projects are targets of anticipation, there is also an aftereffect in which it is difficult to discern what exactly has happened: ecological trauma is just as much a scene of uncertainty as are ecological futures. Especially when subtle environmental effects are hard to pinpoint, or tacit relations are undermined in ways that cannot be identified straightforwardly, the failure of language to grasp ecological change is as pertinent to experience as explicit debates about the constitution of reality. In turn, foregrounding the production of uncertainty does not substitute for an analysis of harm, rather, it adds to it by underlining a crucial element of experience and making clear that these experiences are the outcome of knowledge practices that render certain lives unintelligible. To be a subject of late industrial environments is to live in a world that is to some degree incomprehensible, and that is not simply a natural part of the human condition but an experience specific to the forms of technology and power at work.

This dam is but one modest instance of the historically unprecedented conditions of harm created by late industrial environments. Much of that harm comes from rendering impossible the conditions in which people's lives and loves make sense. The production of uncertainty indicates the value of knowledge to living because it renders existing worlds unintelligible and disables existing, technically sophisticated strategies of living. Late industrial environments thrust wholly new problems and predicaments onto people, such that Lao villagers suffer ruinous attempts at amelioration that never would have happened without the intervention of American activists. Anthropogenic ecologies create not only harmed bodies, but also people capable of bearing harm. The extent of this, too, is historically unprecedented. From the Chernobyl nuclear incident to the pervasive accumulation of industrial toxins, disasters slow and fast make anthropos into unfinished business.

The cocktail of political domination, capitalist appropriation, and ecological and biological transformation has pushed capacities for living into wholly novel contortions.

Povinelli (2011) describes endurance and exhaustion as critical experiences of living in conditions of abandonment. In doing so, she calls attention to the "ordinary, chronic, and cruddy" forms of living and dying, and the work of "suffering . . . , enduring and expiring" in late liberalism (13). Povinelli's work provides a welcome rejoinder to biopolitical analyses that focus exclusively on the productivity of power over life. In contrast to sovereign forms whose limit of power was the ability to kill or let live, Foucault famously argues for "the emergence of a power [that] consists in making live and letting die" (2003, 247). Letting die, and the tactics of living in late industrial environments it provokes, realize an abandonment in which forms of living are unmoored from the conditions that make them possible. This condition of abandonment is an essential dimension of risky late industrial environments, in which pervasive uncertainty demonstrates the extent to which quality knowledge is integral to projects of living. The production of uncertainty does not merely juxtapose confident fact against discursive ambiguity. Rather, to say that the sustainability work was inadequate both demonstrates the long-term, systemic underinvestment in addressing environmental problems and articulates a demand and a judgment on its failed expertise. The production of uncertainty foregrounds a biopolitics of allowing life to die.

Finally, the biopolitical framework of managed care remains critical, but anthropogenic rivers exceed the biopolitical. In this chapter I have emphasized the oscillating interplay of sustainability work and the transformed riparian environment to put forward a language for addressing anthropogenic rivers. In particular, I have been concerned to describe the temporal flux between water and soil and to show that hydropower continually poses the question of what it means to live with rivers through the unstable relation between wet and dry. That riparian topology or aquacline is a scene of living, full of activity and investment. Too much water in the soil, suspended solids in the water, unpredictable flooding, tons of sediment that course down the riverbed—these are ongoing, irreversible transformations that do not settle down but rather stage an event that must be reckoned with for decades. Technological artifacts temporalize, and people work hard to entrain themselves to these unstable conditions of emergence. When is an anthropogenic river? *Indefinitely. Irreversibly.*

Conclusion

Figuring the Anthropogenic

I have argued that the anthropogenic is a double relation in which rivers are transformed through the technological interventions of late industrialism, and yet, reciprocally, these rivers transform capacities for life and living in ways that are pervasive and frequently unexpected. Far from thinking of late industrial hydropower in terms of regularized rivers in which the experts and technicians maintain control, sustainable hydropower formations are pervaded by inadequate techniques, systematic underinvestment in research, and scant knowledge. In a context of massive discursive production, the experts simply do not know how to mitigate hydropower development. This is true even when the dam is commercially successful and widely praised even by some critics for being better than most. It is also true for state-of-the-art "model" World Bank projects. Because "technological artifacts temporalize, opening us to a future that we cannot fully appropriate even as they render us subject to a past that is not of our making" (Braun and Whatmore 2010, xxi), they are thereby constantly surprising, and in that temporal flux anthropogenic rivers make vast, diffuse demands on the extended relationships they organize and make possible. In answering to these demands, people become more than they were.

In my use of the term, "anthropogenic" is a figure that allows attention to several corresponding aspects of late industrial environments. The first is that technological infrastructures dramatically and unpredictably reshape capacities for life and living, and in so doing they both entrain new political socialities and decompose existing formations. Large-scale sociotechnical systems make for long-term investments that thrust political sociality into unexpected domains. They are difficult and contingent achievements that nonetheless constitute the built form of a radically transformed Earth ecology. Hydropower is only one small part of this. But it is impossible now to pretend that modernity has resulted in human rationalization and control over nature. Likewise, it is inadequate to think this anthropogenic influence simply in terms of environmental destruction. An anthropogenic ecology orchestrates oscillating connections between air conditioning in provincial Thailand and the crumbling stream banks of central Laos. An anthropogenic ecology is one in which tropical biodiversity has the capacity to organize vast networks of venture capital and genomic science (Hayden 2003). An anthropogenic ecology is one in which the Aeolian promises of wind and power can constitute a vivid atmospheric sociality (Howe and Boyer 2016). Late industrial environments are not political only because they stimulate controversy or because they result in winners and losers. They are political because they entrain certain capacities for living while rendering others impossible.

Furthermore, anthropogenic experiments turn out in all kinds of unpredictable ways, and that experience is indeed part of the warp and weft of living within late industrial environments. The biopolitics of the anthropogenic hinges on the production of uncertainty. My use of the term "underdetermination" bypasses the twin errors of determinism and indeterminacy. Indeterminacy is inadequate because it emphasizes too easily a sense of purely open contingent possibility when in fact there is a lot to be said for the kinds of relations at stake.[1] Through the deliberate work of a relatively small number of activists, the Lao hydropower industry had to figure out how to see itself in a different way. Underdetermination does not mean that celebrated technical entrepreneurs can pull off any feat they can imagine. Rather, it means that their work on real, historically specific relations contains an imaginative dimension that is oriented to the unexpected potential of those relations. Because material relations are not deterministic, these actors appropriate possibilities that others may not recognize; and their imagination therefore references an ontological potential that is distinct from other ways of imagining those relations. And yet, in that very activity, they inevitably fail to recognize latent potentials that make things turn out differently than expected. Because material relations are real relations requiring work, those actors are obligated to them within the terms of their own projective understanding. It is *their* vernacular materialism at stake in their obligations, not a theoretical materialism dependent on Euro-American philosophy; it is *their* materialism that constitutes locally specific experimental ontologies (Jensen and Morita 2015). Put differently, ontology and epistemology must be thought at the same time, in the same gesture. This ethnography has therefore been an exercise in understanding vernacular materialisms, specifically those that dominate American-influenced practical reason, in which I identify similarities in practice and ethos among managers, technicians, and activists.

Conversely, rivers do not operate anthropogenically through any purely material force relations, nor can the ontology of anthropogenic rivers be separated from the complex knowledge relations that modify them or subvert them for ulterior purposes. Rather, specific persons bear obligations to rivers because of their own designs, their histories of creative reappropriation of ecological possibility, and their own committed materialisms. Insisting that such relations are material-semiotic only goes part way. By focusing on questions of anticipation, promise, or speculation, I have sought to show a temporality in which material relations constantly exceed discursive capacities. There is a matter of seduction at stake in how rivers entrain their humans. Following a Deleuzean reading of Foucault, I have used the term "problematization" to denote the risk and failure in which thinking remains open to the event of ecology. Indeed it is through the risk of their riparian obligations that people have the capacity to think rivers as such. When people talk about sustainable hydropower, they are trying to keep up with the massive flow of ecological change that they barely understand, but they are also seduced by the possibilities at hand. For this reason, the anthropogenic is a creative domain for reimagining and reappropriating late industrial ecologies. If nothing else, this book should stand as an argument for how anthropogenic ecologies so much unexplored potential for diffuse political socialities in the present. We have no idea what kinds of beings we might become.

Thus, the anthropogenic emphasizes the expansive capacities through which we might ask *of what is the human capable?*. Anthropos is at stake not because the human can be distinguished from a purified version of "nature" or because an essential species-identity has somehow come to define the planetary present, but because collective capacities of being and doing constantly take new and unexpected ecological form. The anthropogenic is offered as a way to speak to a specific predicament of the contemporary, not to characterize human ecologies at all places and times or offer a new genetic principle or tale of genesis. The anthropogenic therefore argues for a nonessentialist ontology in which the human is unfinished business.

The Uncertainty of Late Industrial Environments

If anthropogenic is offered as a figure, *uncertainty* should be understood as a concept that attempts to bring specificity to the relations of knowledge and late industrial ecologies. Developing a concept of uncertainty and building on the work of Kim Fortun, Michelle Murphy, Adriana Petryna, and others has been essential to this project because the dominant modalities for thinking about power have been concerned with the constitution of naturalized facts or the reality effect of discursive regimes. The limitations when applied to ecology have only been too apparent, and the literature surrounding late industrial environments suggests that a large range of problems can be clarified by thinking about sites in which the active production of uncertainty is underway. I am certainly not arguing that there are no truth effects in which ecological relations appear naturalized (i.e., their historical and political relations are hidden) or that there is no effort to constitute facts as such. On the contrary, those practices, like the production of uncertainty undertaken by the actors I have described, constantly respond to a problematic situation. Late industrial environments are constituted by uncertainty because they are experienced as intractable problems, unresolved troubles, risky opportunities, and vital threats. Environmental knowledge, like medical knowledge, is subordinated to a domain of the event that dominates and exceeds that knowledge. Uncertainty, I argue, is critical for thinking anthropogenic ecologies biopolitically while holding open the ways postnatural relations exceed the powerful knowledges that help constitute them.

Features of this conceptual approach to uncertainty include: uncertainty is an inherent feature of contemporary knowledge and practical reason, not an absence of certainty or a simple limit to knowledge; uncertainty (like knowledge) is built into the durable relations that constitute our ecological present; and uncertainty demonstrates the value of knowledge. It is important to remember that knowledge is not a thing but an aspect of relationships; separating knowledge out as a mental phenomenon (e.g., belief) or a collective phenomenon (e.g., discourse) always risks obscuring the real relations in which it participates. Let me elaborate these features.

Uncertainty is not the opposite or lack of certainty; rather, it is inherent in the practice of knowledge. Consider this passage concerning testimony from Kant's lectures, discussed by Axel Gelfert, for it demonstrates Kant's willingness to foreground critical faculties in the domain of empirical knowledge: "to be sure, the testimony that we accept from others is subject to just as many hazards as our own experience is subject to errors. But we can just as well have certainty through the testimony of others as through our own experience" (Kant quoted in Gelfert 2006, 633). For Kant, this kind of empirical certainty, whether through testimony or experience, is precisely *never certain* in the way pure reason promised axiomatic certainty.[2] Rather, the testimony of others and personal experience both require a critical attitude as a manner of being familiar with their limitations. As such, "historical belief is reasonable, if it is critical" (quoted in Gelfert 2006, 641). As Colin Koopman has shown, this dimension of critical empirical judgment was later appropriated by Charles Sanders Peirce and became one touchstone of American philosophies of epistemic pluralism—influencing, I have argued, American management theory. "Pragmatic belief [is] such contingent belief, which yet forms the ground for actual employment of means to certain actions" (Kant quoted in Koopman 2009, 113). Gelfert confirms this by underscoring that for Kant critical activity implies competence, which might be read as a formulation of expertise. "Competence eludes formal definition because it is essentially a skill that varies with context: 'It is not in all cases so easy to acquire experiences, and it takes practice' [Kant]" (2006, 631). To summarize, empirical knowledge and historical belief are never certain in any absolute sense—the uncertainty is inherent in what it means to know and think, as it were; they require critical judgement; and that judgment demands acquired competence. Uncertainty is a built-in feature of practical reasoning, empirical knowledge, and the articulated techniques of expertise I have described for Lao hydropower.

Because late industrial ecologies are partly an effect of knowledge practices—design, engineering, or finance among many others—so too the limitations and inadequacies of that knowledge are also part of the anthropogenic landscape. In the most sustained treatment of uncertainty in the anthropological literature, Limor Samimian-Darash and Paul Rabinow argue that risk and uncertainty should be treated "as concepts rather than as things in the world" (2015, 3). The evidence I have provided in this ethnography suggests a different view, that knowledge is an aspect of relations with things in the world, and it is not possible to separate concepts and things as such, for the simple fact that human knowledges have participated in the emergence of that world. If we accept that engineering, design or other human knowledge practices are essential to understanding the built environment (Rabinow 1989; Winner 1980)—or even biological organisms (Rabinow 1996)—then by the same gesture we must accept that the limitations of our knowledge are equally realized. As Giles Deleuze argues, "personal uncertainty is not a doubt foreign to what is happening, but rather an objective structure of the event itself" (1990, 3). By contrast, the analysis presented in Samimian-Darash and Rabinow 2015 is formally agnostic about the actuality of real situations because they treat discourse as an autonomous field of representation.[3] Although it is an improvement, even saying that knowledge (and therefore uncertainty) is a thing in the world misses the point. Rather, the limitation is only overcome once it is recognized that knowledge is a dimension of real, historically specific relationships among people and things, and uncertainty is not incidental to but an integral feature of those relations.

In developing an ethnography of expertise, I have sought to explore what people do when they are obligated to acknowledge how little they know, the kinds of tactics that have arisen when claims of authority or technical control are viewed as a liability, and the practices that emerge when things do not go as planned. Continuing from the discussion of Kant, my understanding of expertise is different from the classic Weberian concern with functionaries who are required to navigate the bureaucratic state, or from assessments that limit expertise to elite authoritative knowledge. By contrast, I view expertise as defined by the actuality of the situation to which knowledge is made to answer and thereby take knowledge *as such* as integral to that actuality. One might say that the expert is she who diagnoses the problematic, which is to say, the one who articulates the normativity of a problematic event. In fact, situated knowledge (Haraway 1988) is precisely that knowledge which articulates the problematic event; it is situated in the space of an event, not in terms of an identity or a heritage or culture. If I have focused here on managers, activists, and consultants, it is not because they have the last say in how we should view anthropogenic rivers, but because their situated knowledges play critical roles in defining the situation at hand, including its vital contingencies.

When it comes to late industrial environments, uncertainty is not just a feature; rather, its role is constitutive. The reason is simple. Because industrial technologies have vast side effects to the technical knowledges that make them possible, they generate deeply problematic situations by constantly denaturalizing the tacit predictability of lived relationships. The extensiveness of the changes they induce is part of the equation. In Isabelle Stengers's language, these novel relations come to *force thought* on people. They constantly undermine workable assumptions. They introduce all sorts of "new things in the world," whether these are locally specific, such as a river that continues to transform itself unpredictably, or whether they have broader significance, such as the methyl isocyanide that permeated bodies on the night of December 2, 1984, and for decades after rendered them vitally different. For Fortun, the pervasive uncertainties of the Bhopal disaster not only undermined existing expertise, but also formed the basis for critical, clinical, and scientifically informed environmental politics. "What counts as expertise has been complicated. Scientific inputs have been crucial to direct diagnostic and therapeutic agendas, but also to validate legal arguments. Science has also been unsettled by the rigorously nonlinear, unpredictable, cumulative effects of toxic chemicals" (2001, xvi). The activity of problematizing an ecological situation plays the central role in my narrative and distinguishes uncertainty as a concept from terms such as "contingency" or "indeterminacy" (which imply open-endedness but not partial knowledge per se). Uncertainty is the constitutive feature of late industrial environments because knowledge is essential to how these environments have come about and because, even after the demise of trust in authoritative expertise, a premium is still placed on knowledgeable expertise for contending with the afterlives of late industrialism.

Late industrial environments are constituted by uncertainty also because partial knowledge is always at stake in the risky, high technology projects that reconfigure the real relations in which life takes place. Things that are not known, systematic ignorances, and problems that are barely acknowledged are a feature of the real relations that come from knowledgeable activity in the world. Put differently, one's capacity to act far exceeds the ability to know how one acts, and it is easy to make a mess of things. But that is only one end of the spectrum, for we can understand only too well that this zone of intractable problems is maintained through all sorts of deliberate practices and likewise forms a substantial part of the domain of activity for powerful actors who are comfortable operating in conditions of uncertainty. Resource developers and managers are *good at* producing messy, unstable environments and continuing to act in their own interests within that chaos; any anthropology of late capitalist environments needs to assess those capacities. Finally, we must be able to specify the systemic destabilization of actual ecologies and pervasive underinvestment in certain kinds of knowledge as two sides of the same dynamic. Knowledge is part of the warp and weft of contemporary political engagement with socionatural conditions. If that is true, then we must also recognize our near-total inability to anticipate and grapple with the diverse repercussions of massive technological achievements.

Notes

Preface

1. This way of formulating the question comes from Rabinow 2003, 2008. See also Rees, forthcoming, *After Ethnos*; Pandian, forthcoming, *A Possible Anthropology*; and Murphy (2008, 696): "In the twenty-first century, humans are chemically transformed beings."

2. As I suggest below, the book can be read against the grain as an anthropology of white masculinity in postcolonial context; also, nearly all of the anthropology of these kinds of capitalist practices that I have found useful comes from feminist authors: Ong 1987; Fortun 2001; Murphy 2006 (a historian); Martin 1994; and Strathern 1992.

Introduction

1. Unindexed recording in author's possession, dated July 26, 2004.

2. See Huet 2012, chap. 9.

3. See Dean 1999, 177; Hacking 1990. For Dean, risk is a way of rendering an objective condition calculable. Hacking is concerned with the techniques of knowledge through which the vagaries of nature came to be understood as probabilistic.

4. On the promissory, see Sunder Rajan 2005; Fortun 2008; Fischer 2009.

5. This problem is directly related to fusing concerns with biopower—including those of hydropower and sustainability—with science studies concerns with the production of knowledge. Among other scholars, Isabelle Stengers's argument concerning obligation and requirement in scientific practice has been an important influence in attempting to delimit this kind of technical entrepreneurialism or speculative materialism as a discrete analytical problem. On this question, Stengers breaks from actor network theory, in spite of her close connection with Latour and the influence his work has had on her. It can be formulated in terms of asymmetry and value, which are absent in Latour until he, very late, determines to consider matters of concern alongside matters of fact (Latour 2004). As Stengers (2010, 52) puts it, "artifacts, operations of mobilization, are *not* all equivalent. Establishing their nonequivalence is a sign of the worth of an experimental practice." That nonequivalence is the outcome of a "topology of requirement and obligation" (52–53) in which requirements are placed on an experimental apparatus, the outcome of which places obligations on the scientist: truth matters, under conditions of relevance established in the practice of science itself. In a comment that must apply to Latour's (1993) polemics against "Moderns" and their epistemology, Stengers points her criticism of equivalence toward "relativist irony" and its "abuse of power" (53). Here is the crucial influence on a discussion of a speculative materialism: under conditions of obligation, "rationality . . . becomes synonymous with risk and challenge" (53). If I may rephrase this gesture—in part to recuperate Georges Canguilhem from Latour's (1993) misreading of his epistemological project of a history of error—any idea capable of truth must also be capable of error, within a definite practice of evaluation (see also Rabinow 1994). That is to say, such experimental material commitments obligate a rationality under conditions of risk. To be committed to one's ideas is dangerous, which in turn is an index of their worth. Uncertainty is therefore not simply the opposite of knowledge but part of a relationship that indicates the value of knowledge.

If I appear to be going too far in extending Stengers's argument to the artful, nonscientific work of sustainability management, she herself would accept such a view: "We can characterize the worth of technological-industrial creations in terms of requirements and obligations. We might even be tempted to extend this mode of characterization to the living. . . . Obligation refers to the fact that a practice imposes upon its *participants* certain risks and challenges that create the value of their activity" (2010, 54–55, her emphasis). So, in very practical terms, environmental activism has created new sets of obligations for hydropower managers, but these obligations are not simply a straightforward repetition of the demands made by either activists or villagers. Rather, managers reconfigure what they can get away with—not as a matter of direct control but of threat or opportunity. Hence those obligations must be understood within the rationality of a hydropower firm that openly committed to dealing with its environmental problems, without the content of those problems having been agreed upon or substantively determined at the outset.

6. In spite of its formal definition, extensive debates over the meaning of "sustainability" demonstrate that the term indicates a problematization rather than a political

program, an ideology or hegemonic institutional form (similar to biodiversity, Lowe 2006; on problematization see Rabinow 2003). Often critiqued for meaning whatever the speaker wants it to mean—for being too vague, too economically driven, or too unscientific—the term "sustainability" is experienced as unsettled, and that is very much a condition of working within its field. Since the value-laden term brings together environmental and economic considerations, the threat of co-optation is unavoidable. Sustainability helps give voice to programs of environmental regulation for actors who cannot imagine environment in the absence of market relations, but it is not reducible to accumulation strategies. Because sustainability is addressed to problems (rather than positive facts or existential necessity), like much environmental practice, it shares many features of clinical medical practice. That is, it is diagnostic; it locates problems and attempts to address them. Thus it is pragmatic, power-laden, and operational. And like commercial medical knowledge, sustainability is opportunistic, for it is heavily capitalized around certain infrastructures of care but not others.

7. For knowledge infrastructure, see Star and Ruhleder 1996; Edwards 2006.

8. Email to LaoFAB email listserv, May 7, 2011.

9. As Cavell 1996 shows with J. L. Austin's performative theory of excuses, mistakes, and accidents. On tragedy, see Carson 2006.

10. On imagination as an ethnographic mode, see Stewart 1996; McLean 2009.

11. Due to Arendt's humanism, these two citations may appear incongruous. While much of actor network theory—foreshadowing the ways Latour has been taken up by new philosophies of ontology—rejects the view of the political subject that concerns Arendt, I note that science and engineering involve *practices* of material engagement, most prominently understood in the figure of the modest witness (Shapin and Schaffer 1985; Haraway 1997), who is the subject of such temporal, material practices of reasoning. In fact, "object-oriented" authors as diverse as Annemarie Mol (2002, 50), Jane Bennett (2010, 45), and Hélène Mialet (2012, 133-4) return to the original scene of cognition and thereby retain concern with the Cartesian problem of the irreducibility of thought. (Somewhat shockingly, Mialet seems to reproduce the mind/body dualism, mainly by implying that mind is all extended body; 2012, 135.) In general, I follow the work of extended cognition (Lave 1988; Star and Ruhleder 1996; Murphy 2013) out of concern for what Rabinow (2003) calls the activity of thinking. Hence, Arendt (1998, 5) captures well my concern with anthropogenic environments in her demand that we "think what we are doing," in her case through work, labor, and politics as categories of worldly activity characterizing "the human condition." I sense a powerful confluence with the project Foucault describes as an "historical ontology of ourselves."

12. Is the river's agency at stake (cf. Mitchell 2002; Latour 1993)? Rather than a question of how to distribute agency among people and things, I treat the anthropogenic as a figure-ground relation or "inversion" (Harvey, Jensen, and Morita 2017, 3) in order to pose questions about how people conform themselves to rivers. My sense is that the metaphor of agency has outlived its usefulness. It functioned best as a tactical reversal or polemic in order to make a critical point (which it does very well), but too frequently it turns into a zero-sum game in which people are not allowed agency if the nonhumans

are to have it or a litmus test for a certain style of representational politics rather than a method that generates insights. My approach to the agency question is to point out modes of valuing nonhuman relations with concepts such as evaluation, affirmation, rejection, commitment, or obligation. The valuing of nonhuman relations makes an obligation on the persons in question, what I call an obligation of worth. There is no reason to assume this is a uniquely human characteristic; Myers 2015 suggests that all living organisms evaluate their environs.

13. Fortun 2012b; Douglas and Wildavsky (1983) 2010; Beck 1992; Barry 2001; Fortun 2001; Adams, Murphy, and Clarke 2009.

14. For arguments about authority and naturalization, see Ferguson 1994; Escobar 1995; Latour 1987; Mitchell 2002. The practices I describe are clearly consonant with neoliberal practices associated with governmentality, the mistrust in centralized expertise, and the predilection for "cheap" government (Rose 1999). Rather than apply a governmentality analytic, here I am concerned with the ontological commitments of certain forms of knowledge associated with business practices, which clearly have their own commitments to skillful negotiation of socionatural relationships, and which make highly partial use of expertise and thereby do not rely on the premises of impartiality and certitude usually associated with institutional and scientific knowledge. One could frame it like this: if neoliberalism as political philosophy involved forms of government that rejected centralized expertise and delegated governmental work to highly partial "private" actors such as NGOs and businesses, then this ethnography is essentially a study of neoliberalism "inside out" (Riles 2000), from the vantage of those businesses to whom governmental practice has been delegated.

15. See also Keck and Sikkink 1998; Khagram, Riker and Sikkink 2002.

16. High 2008, 2014; Singh 2010, 2012; cf. High et al. 2009; Goldman 2005. See comparable debates about participatory development: Kesby 2005; Hickey and Mohan 2004; Cooke and Kothari 2001.

17. On management see Martin 1994; Fortun 2001; Strathern 1992; Murphy 2006; Ong 1987.

18. On conservation see West 2006; and in Laos, Singh 2012; cf. Goldman 2005; on regional affirmations, Singh 2011; Sivaramakrishnan and Agrawal 2003. On hope and improvement see, e.g., Mosse 2005; Li 2007; Hughes 2006; Ferguson 1999.

19. Morita (2016, 117) writes of "an ambiguous place between the land and the sea"; see also Shoemaker, Baird, and Baird 2001; Lahiri-Dutt and Samanta 2007; McLean 2011.

20. Terms such as "threat," "opportunity," "apprehension," "seduction," "fear," or "promise" index the work of imagining ecological possibilities. Technologies extend and intensify the specific underdetermined capacities of nonhuman natures but, within the ontological commitments of the theory I present, all real relations are underdetermined to some degree.

Furthermore, the term "indeterminacy" has been used recently to similar effect. In my view, it is a mistake to imply that the debilitating ecological effects of late industrial technologies are indeterminate. To say that a set of ecological relations could turn out to be otherwise, or that given some intervention or unexpected contingency could be made otherwise, is very different from saying that those effects are causally indeterminate (vague, ambiguous, nonspecific, etc.). The effects of Hurricane Katrina were not vague, even if they were fraught with uncertainty.

21. "Uncertainty" is an increasingly relevant term in anthropology. Eschewing a strong distinction between risk and uncertainty, Ong (2016) describes the productive uncertainty that permeates postgenomics in Asia, especially Singapore, when "radical uncertainties" permeate the new epigenetics, which in turn create new social forms that are themselves permeated with novel uncertainties. In much recent work on uncertainty (especially Samimian-Darash and Rabinow 2015), the term "management" plays a critical function but remains untheorized. If uncertainty is what gets managed, then clearly the verb is a major part of how it should be developed as an analytic. Ong's (2016) focus on the creative, productive elements that come along with the uncertainties of biocapital is parallel to the argument I present.

In a useful move, MacPhail (2010) refers to strategic uncertainty to describe how epidemiologists construct a domain of discursive uncertainty within which their scientific practice is intelligible, demonstrating that uncertainty is not a natural condition of undifferentiated nonknowledge. I would argue, however, that uncertainty is a dimension of real, material-semiotic relations because knowledge and its inadequacies are built into anthropogenic environments; see Whitington 2013. Samimian-Darash (2013) also argues that uncertainty is distinct from risk with respect to epidemiological events and locates what she calls "potential uncertainty" in Gilles Deleuze's work on the objective conditions of events. My approach foregrounds the dimension of value or worth and argues that uncertainty is a feature of ecological relations themselves, including epidemiological events. In Samimian-Darash's terms, an "event" is precisely what establishes "a virtual domain with the capacity to generate a broad variety of actualizations" (2013, 4). See further discussion of Samimian-Darash and Rabinow's (2015) understanding of uncertainty in the conclusion.

1. Hydropower's Circle of Influence

1. There is a substantial literature on postsocialism as an anthropological and methodological category (Verdery 1996; Humphreys 2002; the many contributions to Hann 2002); and for Southeast Asia (e.g., Harms 2016; Evans 1998) and Laos in particular (in addition to the many citations in this chapter, see Chua 2012; Evans 1995; Pholsena 2006; and contributions to Bouté and Pholsena 2017).

2. See Whitington 2012 for more detailed arguments.

3. Rigg 2009 has a detailed comparison of the New Economic Mechanism with the Washington Consensus. Suyavong 1997 has a different description of IMF structural adjustment reforms. Boupha 2002 describes the process from the vantage of Laos's governing elite.

4. Baird and Quastel (2015, 1235) provide an extensive discussion of the other side of this story by describing the minimal "regulation by contract" by which environmental oversight "only featured the unilateral assertion of impacts and solutions" of the developers. See also Johns (2015) who discusses modular law for neoliberal Lao hydropower as a program of "failing forward."

5. It is commonly cited that 80 percent of the country's intellectuals, technicians, and functionaries fled the country when the communists came to power (e.g., Vorapeth 2007, 13), throwing into relief both the tremendous role of knowledge in the work of government and the problematic of postrevolutionary rule for a country formed in the crucible of war.

6. Email to LaoFAB listserv dated April 20, 2011.

7. Email to LaoFAB listserv dated May 9, 2011. Acronyms were removed.

8. Email to LaoFAB listserv dated April 20, 2011.

9. Two exceptions that prove the rule stand out. The Houay Ho project built and financed by South Korea was a dismal financial failure. The much smaller Laksao Dam, financed by the semiautonomous military timber enterprise BPKP (Borisat Phatthana Khet Phoudoi), collapsed during a tropical depression, precipitating a downstream disaster.

10. Social and political space has been a concern of French scholarship of Laos, see Condominas 1980; Taillard 1989; Bountavy and Taillard 2000, and is an important theme in the anthropology of Thailand, see, e.g., Tooker 1996; Elinoff 2016.

11. Stuart-Fox's diagnosis of clientele-based corruption in the Lao state stands in stark contrast to his nuanced historiography of continuities and discontinuities in Lao history, Stuart-Fox 1993, and is limited because it confuses a political diagnosis motivated by the concerns of development institutions with construal of a social fact in need of careful elucidation.

2. Vulnerable at Every Joint

1. The claim neglects international climate change negotiations. There is a vast literature on "transnational advocacy networks," some of it by anthropologists and much of it taking a "new institutionalist" international relations perspective broadly compatible with a practice-based approach in social theory. Useful outcomes of this perspective include the willingness to consider that small policy changes can have substantial effects, to allow for institutional heterogeneity, and to pay attention to the sometimes unexpected roles played by heterodox practitioners (Fox 2000, Hochstetler and Keck 2007).

2. IRN was renamed International Rivers after the events I describe here. I retain the original name for consistency.

3. Part of what motivated my approach was work in legal anthropology in which the powers of the law cannot be taken for granted (Geertz 1983: chap. 8; Moore 2001; Obarrio 2014; and Johns 2015); precisely because the theory of force implicit in the law, if one does not accept a theory of the state in terms of a monopoly of violence, remains vague from an ontological standpoint.

4. For important work on the network as native category, see Strathern 1992; Riles 2000.

5. Before Shoemaker's report, there was another activist campaign organized by the Norwegian NGO FIVAS, which prompted an additional consulting report by the project developers and indeed led to partial adaptation of some environmental measures.

6. Unindexed archive, International Rivers, accessed by author August 31, 2006.

7. To foreground the materiality of these practices, consider this hand-written note, on a printed-out email: "-> Treasury—delaying report due to disagreements, concerned ADB is going to water it down—needs to come out quickly. ED [ADB Executive Director] to make inquiries re rel'ship b/t [between] consultant & [company], what's happening etc." Unindexed archive, International Rivers, accessed August 31, 2006.

8. As in this request: "1) Can you tell me if the license agreement between the government and concessionaire is a public document? 2) Which institution is given the mandate to license such an agreement? Is it usually the Norwegian WR [water resources] and Energy Admin or is it others? 3) Does the Developer pay for the Environmental Impact Assessment? For the hearing process?" Unindexed archive, International Rivers, accessed August 31, 2006.

9. By contrast, in contemporary anthropological usage, "politics" usually refers merely to relations in which power is at stake. To the extent that social relations are held to be imbued with power, this understanding of politics cannot function as a concept but can only serve to emphasize the power relations at stake. In the commonplace anthropological sense, the claim that "hydropower is political" would simply imply that power relations are involved, that dams are not merely technical, that dams are contested, or that there are winners and losers. There is nothing wrong with these sorts of claims—all certainly apply to the case at hand—but they say little when it comes to matters of ontology. For instance, such claims miss the mainstream politics in which postcolonial nationalism was predicated on large dams, emphasizing instead that the standing of certain national subjects was sacrificed in the process.

10. ADB 1994 Summary Environmental Impact Assessment, Loan 1329-Lao, cover letter. Unindexed archive, International Rivers, accessed August 31, 2006.

11. In a similar way, Adam Fleishmann (2015) argues for a figure of "technicians of the possible" in comparison to Paul Rabinow's two figures of modern expertise. In *French Modern: Norms and Forms of the Social Environment*, Rabinow (1989) describes a particular figure of modern social government who was responsible for implementing general ideas by translating them into locally workable formats. These "technicians of general ideas" were (and are) essential to rolling-out of homogenizing, generalizable social instruments, and stood in contrast to "specific intellectuals" who dedicated considerable intellectual effort to rethinking definite problems of limited scope, often with important results for how others were able to think about analogous problems.

12. As I finalized revisions on this manuscript, a dam under construction in Laos collapsed, releasing over a few minutes an estimated five hundred thousand cubic meters of water into a tiny stream. The video is dramatic, to say the least.

Interlude: Intimacy (Vetting)

1. 2003 BNL4 Contract Addendum: Village Redevelopment. Final Status Report. (Government Document). Vientiane: Electricite du Lao. Page 6. Document in author's possession.

2. Oxford English Dictionary, OED Online (2015), s.v. "vet."

3. Performance-Based Management

1. The sentence reads, "a. Before leaving Laos, the review panel will prepare a draft report and issue such report directly to THPC & IRN before departing Laos." "Terms of Reference, Review of the Environmental Management Division," anonymous undated document in author's possession, p. 5.

2. This is why I write that people are *obligated to* environments, rather than writing that environments obligate people. The obligation is an affirmation.

3. Reflection on the term "management" suggests that the practical grammar of management is highly significant for what is meant by uncertainty. In general, it is possible to group these intractable problems for management in four general areas, namely *repeated operations* (managing material stresses or fatigue or, for instance, ergonomics for carpel tunnel); *large quantities* (traffic, sewage, pain, information); *uncertain variables* (risks, threats, opportunities; operating conditions of all kinds); and *delicate situations* (ecological relations, catastrophic events, or any dynamic requiring finesse). The groupings are not exhaustive categories but generic mnemonics that serve to outline one materialist ontology of postnatural relations.

4. Note this is a completely different etymology than the French *ménage* (household), from the Latin *manēre* (to stay), although OED notes historical "confusion" between the two. Oxford English Dictionary, *OED Online* (2015), s.v. "manage." Oxford English Dictionary, *OED Online* (2015), s.v. "Ménage."

5. The phrase comes from an interview with a private sector manager for climate change adaptation in Miami Beach. Another choice turn of phrase that captures this attitude very well is, "We're not hiring a consultant to hire a consultant" (in Weiss 2016).

6. "I expect a lot from EMD [Environmental Management Division] staff," he told me, "Some people are, like, 'wink-wink, you know the Lao staff,' and never expect very much from them."

4. The Ethics of Document Engineering

1. On ethics, see Faubion 2011; Foucault 1985, 1988; MacIntyre 1984; Rabinow and Bennett 2012.

2. Foucault identifies "practices;" a Stengers-inspired nudge can shift the emphasis to "relations" as I have done.

3. On technical criticism, see, for example, Collier and Ong 2005; Miyazaki 2013, 11; Rabinow 2003, 33.

4. Nelson 2002 demonstrates that transnational environmental NGOs have proven capable of entrenching themselves in multilateral funding mechanisms, but they have met serious restraints in both monitoring concessions made by multilaterals and maintaining the immense volume of monitoring activity necessary for ongoing campaigns. On issues pertaining to development consulting and expertise see, for example, Ferguson 1994; Crush 1995; Li 1999, 2007; Kothari 2005; Mitchell 2002; and Hirsch 2001.

5. On documents, see, for example, Heatherington 2011; Riles 2000, 2006; Harper 1998; Miyazaki 2013; Strathern 2000; Gupta 2012; and Rottenburg 2009.

6. The analysis presented in this section is based on two versions of a draft document dated March 2004, in the author's possession, including a draft written by the consultants Steve, Umporn and Douglas discussed in this chapter and a version of this same document that contains tracked editorial changes and commentary made by a manager of THPC. These two draft versions have been cross referenced against a final version of the document. The documents cannot be fully referenced for confidentiality reasons.

7. The only rule of thumb made explicit was that if, in general, 10 percent of the development activities attempted had a lasting significance, then the project overall could be considered a success. Ten percent is not much—especially when the issue is repairing a damaged livelihood.

8. While there is no concern with ecology in any of Foucault's work, nearly all of his work was preoccupied with thought in relation to the historical formation of nature, from his concern with the science of madness to the architecture of natural classification to the experience of desire. The tension between the problematic and the event also identifies ontology as a kind of metaphysics (Skafish 2014; Charbonnier, Salmon, and Skafish 2017), since the reality of an event depends in part on an "affirmation" (Foucault 1998, 359): "it [the problem] seriously disobeys the Hegelian negative because it is a multiple affirmation; it is not subjected to the contradiction of being and non-being, since it is being. We must think problematically rather than question and answer dialectically." Notably, this implies that all substantialisms are metaphysical, whether concerning an ethnographic metaphysics that prioritizes the reality of cultures, practices, or meanings; the naïve realism of scientific positivism that obscures its metaphysical commitments to visibility as well as its own historical judgment about what is real and what is fantasy; the object oriented ontology that refuses to examine the intrinsically historical and relational status of scientific knowledge; or the ontology of information systems that requires a formal typology of digital entities and ordering systems much as described by Geoff Bowker and Susan Leigh Star (2001). This is partly why I am skeptical of any theory of ontology that does not explicitly situate itself within an intellectual genealogy as a matter of reflexive critique. Claims to radically break with the historical conditions of one's thought are inevitably misleading. Hence my concern with an anthropology of postnatural experimental relations rather than a disavowal of the nature-society distinction.

5. Anthropogenic Rivers

1. The word for "nature," *thammasat*, is formed around the root word for dharma. Tanaka (1993, 133) reports from his ethnography of Lao socialist agronomy: "Sometimes they unconsciously expressed a feeling of shame by laughing at themselves for the backwardness of their techniques. They often used the word thammasat self-contemptuously."

2. The contrasting political values of *phatthana* and *chaloen* have been discussed by Thongchai 2000; Creak 2010; Johnson 2013; Whitington 2014. Thanks to Nick Enfield for discussion on this point.

3. Intensification is a Deleuzean metaphor for describing anomalous, borderline "multiplicities," or agglomerations whose distinctiveness comes from relations that multiply and feed off of each other in a strange attraction, so that no component of the multiplicity can exist without the other (Deleuze and Guattari 1987, chap. 10).

4. "Project Design Checklist." Anonymous, undated document in author's possession.

5. As I suggested in chapter 3, this should be understood as a kind of hypothetically self-correcting, cybernetic feedback logic, like that developed for missile tracking of aircraft through infrared sensing that constantly self-corrects rather than trying to predict the target once and for all at the outset. Similarly, Dominic Boyer discusses Clifford Geertz's interest in cybernetics as proposing "a model of culture not as concrete behavior patterns but rather as a 'set of control mechanisms—plans, recipes, rules, instructions (what computer engineers call 'programs')—for the governing of behavior'" (Boyer 2015, 536 quoting Geertz 1973, 44).

6. I greatly appreciate two books that develop similar kinds of analysis: Rottenburg 2009 and Jensen and Winthereik 2013. Thanks to an anonymous reviewer for recommending them.

7. A striking iteration of this is a series of self-help tutorials by Maxwell Maltz, a plastic surgeon, from the 1960s to the 1970s in the United States under the rubric of "psycho-cybernetics" (e.g., Maltz 1969). Part of the cognitive behaviorism movement, psycho-cybernetics counselled that therapeutics should not be concerned with the origins or causes of psychological experience but with achieving results in creating a new psyche. The central technique is that the subject can develop an image of whom he or she wishes to become and then, holding that image in view as the center point of a daily practice of self-reference, cybernetically work to perform that new self through positive feedback. Contrary to viewing this as a purely mental technique, visualization is meant to create an integrated feedback linking mind, body, and performance. Needless to say, it is resolutely future-oriented: the image (output, achievement, telos) is the signified of a system that explicitly declares disinterest in historical causes or the meaning of one's past experience. The achievement is one of exteriority—getting outside one's head through a specifically image-based performance of self.

This kind of psychic manipulation problematizes the volitional subject of sovereign rights, choice, and decision making within a genealogy of management. Jonathan Crary would recognize it as a latterly instance within a history of attention and perception: "that human subjects have determinate psychophysiological capacities and functions that might be subject to management (regardless of the effectiveness of such strategies) has been the underpinning of institutional strategies and practices for over a hundred years" (1999, 76). While psycho-cybernetics appears to celebrate a free subject who choses her future self, the visualization technique and cybernetic metaphor implement a system of exteriority and subjective manipulation. When Berlant writes of a "melodramatic view of individual agency [that casts] the human as most fully itself when assuming the spectacular posture of performative action" (2011, 96), I only want to ensure we understand that view of agency as highly subjected and self-subjecting. Such a vision of agency is one that takes on outsized and impossible obligations (risks). It is not the classical subject of free choice, rational knowledge, and decision, but a neoliberal subject of relational obligation, pervasive self-surveillance, and contingent manipulation. Concerning late industrial environments, the fact that stakeholder roundtables are viewed as antipolitical and manipulative and bear an uncanny resemblance to anthropological techniques of understanding qualitative relations (Fortun 2001; Ottinger 2008), shows the need for better understanding of the diversity of relational knowledges. As Strathern (2005) reminds us, not all understandings of the relation are equivalent.

8. Unindexed recording in author's possession, dated July 26, 2004.

9. "Mitigation and Compensation Program," anonymous report, Sept 7, 2000. Theun-Hinboun Power Company, Vientiane, Laos.

10. Baird and Quastel (2015) also describe the intimate connection among Thai air conditioning use, electricity prices, and ecological damage along Lao rivers, with a focus on policy concerns for a political ecology of grid management.

Conclusion

1. For a prominent recent emphasis on indeterminacy, see Tsing 2015; to her credit, she is concerned with diverse knowledge relations and temporalization; in her words, "What if our indeterminate life form was not the shape of our bodies but the shape of our motions over time" (47). I fear only that the term "indeterminacy" overly celebrates the vagueness of the encounter and the momentary.

2. Whether axiomatic proof ever achieves truth is a wholly separate matter.

3. While there is much to be considered in their volume, the shift they identify to the management of uncertainty from the control of risk is broadly consonant with my claims. I would have wished to see critical reflection on the term "management" in their text, since the verb form of the practical operation is what ultimately must define the meaning of uncertainty. Thus they treat uncertainty as a species of purely discursive expert reasoning while declining to include problematical situations into their concept work; world and concept are formally separated. Hence, implicitly, they also distinguish between serious speech acts and the broader sorts of vernacular knowledges that might provide a richer examination of the contemporary problematic (see Dreyfus and Rabinow 1983). For this reason, they miss the broader semantic resources of the term "management," and in claiming that uncertainty is what gets managed, they offer no reflection on the significance of the term. "Management"—as practice and metaphor—I have argued, is actually essential to how uncertainty has gained conceptual currency.

Such a view that uncertainty is a feature of contemporary ecologies refocuses discussion onto ecological events specific to late industrial environments. Concern for ecological events shows that uncertainty characterizes not only the forms of knowledge at stake but also the ecologies those knowledges address, such as climate change or emergent pandemic viruses. In other words, pandemic diseases already exist in potentia—as real potentials that exist separate from any expert discourse—and uncertainty describes our limited, expert relation to that potential, not merely the contours of an increasingly savvy form of expertise. Their expert informers engage with, judge, and affirm the diseases as ontological potentials, but they do not.

Bibliography

Abu el-Haj, Nadia. 2002. *Facts on the Ground: Archaeological Practice and Territorial Self-Fashioning in Israeli Society*. Chicago: University of Chicago Press.

Adams, Vincanne, Michelle Murphy, and Adele E. Clarke. 2009. "Anticipation: Technoscience, Life, Affect, Temporality." *Subjectivity* 28 (1): 246–65. doi:10.1057/sub.2009.18.

ADB. 1997. "Hydropower Project to Increase Lao PDR's GDP by 7 Percent." *Asian Development Bank Review* (Manila), December, page 8.

——. 1998. Report on Site Visit 6–9 May 1998, Loan 1239-Lao (SF): Theun-Hinboun Hydropower Project, Special Loan Review Mission. Manila: Asian Development Bank.

——. 2001. Participatory Poverty Assessment. Vientiane: Asian Development Bank.

Arendt, Hannah. (1958) 1998. *The Human Condition*. 2nd ed. Chicago: University of Chicago Press.

Baird, Ian G. 2011. "Turning Land into Capital, Turning People into Labour: Primitive Accumulation and the Arrival of Large-Scale Economic Land Concessions in the Lao People's Democratic Republic." *New Proposals: Journal of Marxism and Interdisciplinary Inquiry* 5 (1): 10–26.

——. 2014. "Political Memories of Conflict, Economic Land Concessions, and Political Landscapes in the Lao People's Democratic Republic." *Geoforum* 52 (March): 61–69. doi:10.1016/j.geoforum.2013.12.012.

Baird, Ian G., and Noah Quastel. 2015. "Rescaling and Reordering Nature–Society Relations: The Nam Theun 2 Hydropower Dam and Laos–Thailand Electricity Networks." *Annals of the Association of American Geographers* 105 (6): 1221–39. doi:10.1080/00045608.2015.1064511.

Baird, Ian G., and Bruce Shoemaker. 2007. "Unsettling Experiences: Internal Resettlement and International Aid Agencies in Laos." *Development and Change* 38 (5): 865–88.

Baird, Ian G., Bruce P. Shoemaker, and Kanokwan Manorom. 2015. "The People and Their River, the World Bank and Its Dam: Revisiting the Xe Bang Fai River in Laos: The People and Their River Revisited." *Development and Change* 46 (5): 1080–1105. doi:10.1111/dech.12186.

Banerjee, S. B. 2003. "Who Sustains Whose Development? Sustainable Development and the Reinvention of Nature." *Organization Studies* 24 (1): 143–80. doi:10.1177/0170840603024001341.

Barad, Karen. 2003. "Posthumanist Performativity: Toward an Understanding of How Matter Comes to Matter." *Signs: Journal of Women in Culture and Society* 28 (3): 801–31.

Barney, Keith. 2014. "Ecological Knowledge and the Making of Plantation Concession Territories in Southern Laos." *Conservation and Society* 12 (4): 352. doi:10.4103/0972-4923.155579.

Barry, Andrew. 2001. *Political Machines: Governing a Technological Society*. London: Athlone Press.

Baviskar, Amita. 2005. *In the Belly of the River: Tribal Conflicts over Development in the Narmada Valley*. 2nd ed. New Dehli: Oxford University Press.

Beck, Ulrich. 1992. *Risk Society: Towards a New Modernity*. Newbury Park, CA: Sage Publications.

——. 1999. *World Risk Society*. Malden, MA: Polity Press.

Bennett, Jane. 2010. "Thing-Power." In *Political Matter: Technoscience, Democracy, and Public Life*, edited by Bruce Braun and Sarah Whatmore, 35–62. Minneapolis: University of Minnesota Press.

Benson, Peter, and Stuart Kirsch. 2010. "Corporate Oxymorons." *Dialectical Anthropology* 34 (1): 45–48.

Berlant, Lauren. 2011. *Cruel Optimism*. Durham, NC: Duke University Press.

Berman, Anders. 2017. "The Political Ontology of Climate Change: Moral Meteorology, Climate Justice, and the Coloniality of Reality in the Bolivian Andes." *Journal of Political Ecology* 24: 921–38.

Biggs, David. 2006. "Reclamation Nations: The U.S. Bureau of Reclamation's Role in Water Management and Nation Building in the Mekong Valley, 1945–1975." *Comparative Technology Transfer and Society* 4 (3): 225–46. doi:10.1353/ctt.2007.0001.

——. 2010. *Quagmire: Nation-Building and Nature in the Mekong Delta*. Seattle: University of Washington Press.

——. 2012. "Small Machines in the Garden: Motorized Pumps and a Technological Revolution in the Mekong Delta." *Modern Asian Studies* 46 (1): 47–70.

Blaser, Mario. 2013. "Ontological Conflicts and the Stories of Peoples in Spite of Europe: Towards a Conversation on Political Ontology." *Current Anthropology* 54 (5): 547–68.

Bountavy, Sisouphanthong, and Christian Taillard. 2000. "Atlas of Laos: The Spatial Structures of Economic and Social Development of the Lao People's Democratic Republic." Copenhagen, Denmark: Nordic Institute for Asian Studies.

Boupha, Phongsavath. 2002. *The Evolution of the Lao State.* Delhi: Konark.

Bouté, Vanina, and Vatthana Pholsena, eds. 2017. *Changing Lives in Laos: Society, Politics and Culture in a Post-Socialist State.* Singapore: National University of Singapore Press.

Bowker, Geoff, and Susan Leigh Star. 2001. *Sorting Things Out: Classification and Its Consequences.* Cambridge, MA: MIT Press.

Boyer, Dominic. 2015. "Anthropology Electric." *Cultural Anthropology* 30 (4): 531–39. doi:10.14506/ca30.4.02.

Braun, Bruce, and Sarah Whatmore, eds. 2010. "The Stuff of Politics: An Introduction." In *Political Matter: Technoscience, Democracy, and Public Life*, ix–xxxviii. Minneapolis: University of Minnesota Press.

Buell, Lawrence. 2011. "Forward." In *Environmental Criticism for the Twenty-First Century*, edited by Stephanie LeMenager, Teresa Shewry, and Ken Hiltner, xiii-xvii. New York: Routledge.

Butler, Judith. 1990. *Gender Trouble: Feminism and the Subversion of Identity.* New York: Routledge.

———. 1993. *Bodies that Matter: On the Discursive Limits of "Sex."* New York: Routledge.

Callon, Michel, Pierre Lascoumes, and Yannick Barthe. 2011. *Acting in an Uncertain World: An Essay on Technical Democracy.* Cambridge, MA: MIT Press.

Candea, Matei. 2014. "The Ontology of the Political Turn." Theorizing the Contemporary, Cultural Anthropology (website). Accessed January 13, 2014. https://culanth.org/fieldsights/469-the-ontology-of-the-political-turn.

Canguilhem, Georges. 1989. *The Normal and the Pathological.* New York: Zone Books.

Carson, Anne. 2006. *Grief Lessons: Four Plays by Euripides.* New York: New York Review of Books.

Cavell, Stanley. 1996. *A Pitch of Philosophy: Autobiographical Exercises.* Cambridge, MA: Harvard University Press.

CGIAR. 2013. "Mekong Hydropower Map and Portal." Digital map database. CGIAR (website). Accessed April 1, 2014. https://wle-mekong.cgiar.org/maps/.

Chambers, Robert. 1974. Managing rural development. *Institute of Development Studies Bulletin* 6 (1): 4–12.

Charbonnier, Pierre, Gildas Salmon, and Peter Skafish, eds. 2017. "Introduction." In *Comparative Metaphysics: Ontology after Anthropology*, 1–24. London: Rowan & Littlefield International.

Chua, Liana. 2012. *Southeast Asian Perspectives on Power.* New York: Routledge.

Cohen, Lawrence. 2008. "Science, Politics, and Dancing Boys: Propositions and Accounts." *Parallax* 14 (3): 35–47. doi:10.1080/13534640802159112.

Collier, Stephen J., and Aihwa Ong. 2005. "Global Politics, Anthropological Problems." In *Global Assemblages: Technology, Politics, and Ethics as Anthropological Problems*, edited by Aihwa Ong and Stephen J. Collier, 3–21. Malden, MA: Blackwell Publishing.

Condominas, Georges. 1980. *L'espace Social a Propos de L'Asie Du Sud-Est*. Paris: Flammarion.

Cooke, Bill, and Uma Kothari, eds. 2001. *Participation: The New Tyranny?* London: Zed Books.

Crary, Jonathan. 1999. *Suspensions of Perception: Attention, Spectacle, and Modern Culture*. Cambridge, MA: MIT Press.

Creak, Simon. 2010. "Sport and the Theatrics of Power in a Postcolonial State: The National Games of 1960s Laos." *Asian Studies Review* 34 (2): 191–210.

Crush, Jonathan, ed. 1995. "Introduction: Imagining Development." In *Power of Development*, 1–26. London: Routledge.

Cruz-Del Rosario, Teresita. 2014. *The State and the Advocate: Case Studies on Development Policy in Asia*. New York: Routledge.

Daston, Lorraine. 2004. *Things that Talk: Object Lessons from Art and Science*. New York: Zone Books.

Davies, Thom, and Abel Polese. 2015. "Informality and Survival in Ukraine's Nuclear Landscape: Living with the Risks of Chernobyl." *Journal of Eurasian Studies* 6 (1): 34–45.

Dean, Mitchell. 1999. *Governmentality: Power and Rule in Modern Society*. London: Sage.

Deleuze, Gilles. 1990. *The Logic of Sense*. Translated by Mark Lester with Charles Stivale, and edited by Constantin V. Boundas. New York: Columbia University Press. Originally published as *Logique du sens*. 1969. Les Éditions de Minuit.

Deleuze, Gilles, and Félix Guattari. 1987. *A Thousand Plateaus: Capitalism and Schizophrenia*. Minneapolis: University of Minnesota Press.

Douglas, Mary, and Aaron Wildavsky. (1983) 2010. *Risk and Culture: An Essay on the Selection of Technological and Environmental Dangers*. Berkeley: University of California Press. First paperback printing, 1983.

Dreyfus, Hubert L., and Paul Rabinow. 1983. *Michel Foucault, beyond Structuralism and Hermeneutics*. Chicago: University of Chicago Press.

Drucker, Peter F. 1954. *The Practice of Management*. New York: HarperCollins.

Dwyer, Michael B. 2013. "Building the Politics Machine: Tools for 'Resolving' the Global Land Grab." *Development and Change* 44 (2): 309–33. doi:10.1111/dech.12014.

———. 2014a. "Micro-Geopolitics: Capitalising Security in Laos's Golden Quadrangle." *Geopolitics* 19 (2): 377–405. doi:10.1080/14650045.2013.780033.

———. 2014b. "Mitigation Expert." In *Figures of Southeast Asian Modernity*, edited by Joshua Barker, Erik Harms, and Johan Lindquist, 99–101. Honolulu: University of Hawai'i Press.

Edwards, Paul. 2006. "Meteorology as Infrastructural Globalism." *Osiris* 21 (1): 229–50.

Elinoff, Eli. 2016. "A House Is More than a House: Aesthetic Politics in a Northeastern Thai Railway Settlement." *Journal of the Royal Anthropological Institute* 22 (3): 610–32.

Escobar, Arturo. 1988. "Power and Visibility: Development and the Invention and Management of the Third World." *Cultural Anthropology* 3 (4): 428–43. doi:10.1525/can.1988.3.4.02a00060.

———. 1995. *Encountering Development: The Making and Unmaking of the Third World.* Princeton, NJ: Princeton University Press.
Evans, Grant. 1995. *Lao Peasants under Socialism and Post-Socialism.* Chiang Mai, Thailand: Silkworm Books.
———. 1998. *The Politics of Ritual and Remembrance: Laos since 1975.* Chiang Mai, Thailand: Silkworm Books.
Farmer, Paul. 1992. *AIDS and Accusation: Haiti and the Geography of Blame.* Berkeley: University of California Press.
Faubion, James D. 2011. *An Anthropology of Ethics.* Cambridge: Cambridge University Press.
Fearnley, Lyle. 2005. "'From Chaos to Controlled Disorder': Syndromic Surveillance, Bioweapons, and the Pathological Future. Report I." Laboratory for the Anthropology of the Contemporary Working Paper No. 5 (website). Accessed April 11, 2018. http://anthropos-lab.net/wp/publications/2007/08/workingpaperno5.pdf.
Ferguson, James. 1994. *The Anti-Politics Machine: "Development," Depoliticization, and Bureaucratic Power in Lesotho.* Minneapolis: University of Minnesota Press.
———. 1999. *Expectations of Modernity: Myths and Meanings of Urban Life on the Zambian Copperbelt.* Perspectives on Southern Africa 57. Berkeley: University of California Press.
———. 2006. *Global Shadows: Africa in the Neoliberal World Order.* Durham, NC: Duke University Press.
Fernea, Robert A. 1969. "Land Reform and Ecology in Postrevolutionary Iraq." *Economic Development and Cultural Change* 17 (3): 356–81.
Ferrie, Jared. 2010. "Laos Turns to Hydropower to Be 'Asia's Battery.'" *Christian Science Monitor*, July 2. http://www.csmonitor.com/World/Asia-Pacific/2010/0702/Laos-turns-to-hydropower-to-be-Asia-s-battery.
Fischer, Michael M. J. 2009. *Anthropological Futures.* Experimental Futures : Technological Lives, Scientific Arts, Anthropological Voices. Durham, NC: Duke University Press.
Fleishmann, Adam. 2015. "A Fertile Abyss: An Anthropology between Climate Change Science and Action." MA thesis, McGill University.
Fortun, Kim. 2001. *Advocacy after Bhopal: Environmentalism, Disaster, New Global Orders.* Chicago: University of Chicago Press.
———. 2003. "Ethnography in/of/as Open Systems." *Reviews in Anthropology* 32 (2): 171–90. doi:10.1080/00988150390197695.
———. 2012a. "Biopolitics and Informating Environmentalism." In *Lively Capital: Biotechnologies, Ethics, and Governance in Global Markets*, edited by Kaushik Sunder Rajan, 306–28. Experimental Futures. Durham, NC: Duke University Press.
———. 2012b. "Ethnography in Late Industrialism." *Cultural Anthropology* 27 (3): 446–64. doi:10.1111/j.1548-1360.2012.01153.x.
Fortun, Michael. 2008. *Promising Genomics: Iceland and deCODE Genetics in a World of Speculation.* Berkeley: University of California Press.
Foucault, Michel. 1985. *The Care of the Self.* Vol. 3 of *The History of Sexuality.* New York: Vintage Books.

——. 1988. *Madness and Civilization: A History of Insanity in the Age of Reason*. New York: Vintage Books, Random House.
——. 1990. "A Concern for Truth." In *Politics, Philosophy, Culture: Interviews and Other Writings 1977–1984*, 255–70. London: Routledge.
——. 1997a. "On the Genealogy of Ethics: An Overview of Work in Progress." In *Ethics: Subjectivity and Truth*, edited by Paul Rabinow, 253–80. Vol. 1 of *Essential Works of Foucault 1954–1984*. London: Penguin.
——. 1997b. "Preface to *The History of Sexuality*, Volume II." In *Ethics: Subjectivity and Truth*, edited by Paul Rabinow, 199–206. Vol. 1 of *Essential Works of Foucault 1954–1984*. London: Penguin.
——. 1997c. "What Is Enlightenment?" In *Ethics: Subjectivity and Truth*, edited by Paul Rabinow, 303–20. Vol. 1 of *Essential Works of Foucault 1954–1984*. London: Penguin.
——. 1998. "Theatrum Philosophicum." In *Aesthetics, Method, and Epistemology*, edited by James D. Faubion. 343–68. Vol. 2 of *Essential Works of Foucault 1954–1984*. New York: New Press.
——. 2003. *Society Must Be Defended: Lectures at the Collège de France, 1975–76*, edited by Mauro Bertani and Alessandro Fontana. New York: Picador. First Picador paperback ed.
——. 2005. *The Hermeneutics of the Subject: Lectures at the Collège de France, 1981–1982*. 1st ed. Lectures at the Collège de France. New York: Picador.
——. 2008. *The Birth of Biopolitics: Lectures at the Collège de France, 1978–79*, edited by Michel Senellart. Basingstoke, UK: Palgrave Macmillan.
Fox, Jonathan A. 2000. "The World Bank Inspection Panel: Lessons from the First Five Years." *Global Governance* 6 (3): 279–318.
Frickel, S., S. Gibbon, J. Howard, J. Kempner, G. Ottinger, and D. J. Hess. 2010. "Undone Science: Charting Social Movement and Civil Society Challenges to Research Agenda Setting." *Science, Technology & Human Values* 35 (4): 444–73. doi:10.1177/0162243909345836.
Funtowicz, S.O., and J. R. Ravetz. 1993. "Science for the Post-Normal Age." *Futures* 25 (7): 739–55.
Geertz, Clifford. 1973. *The Interpretation of Cultures: Selected Essays*. New York: Basic Books.
——. 1983. *Local Knowledge: Further Essays in Interpretive Anthropology*. New York: Basic Books.
Gelfert, Axel. 2006. "Kant on Testimony." *British Journal for the History of Philosophy* 14 (4): 627–52. doi:10.1080/09608780600965226.
Goldman, Michael. 2005. *Imperial Nature: The World Bank and Struggles for Justice in the Age of Globalization*. New Haven, CT: Yale University Press.
Gupta, Akhil. 2012. *Red Tape: Bureaucracy, Structural Violence, and Poverty in India*. Durham, NC: Duke University Press.
Guyer, J. I., N. Khan, J. Obarrio, C. Bledsoe, J. Chu, S. Bachir Diagne, K. Hart, et al. 2010. "Introduction: Number as Inventive Frontier." *Anthropological Theory* 10 (1–2): 36–61. doi:10.1177/1463499610365388.
Hacking, Ian. 1990. *The Taming of Chance*. Cambridge: Cambridge University Press.

Halpern, Orit. 2014. *Beautiful Data: A History of Vision and Reason since 1945*. Experimental Futures. Durham, NC: Duke University Press.

Hancock, Graham. 1989. *Lords of Poverty: The Power, Prestige, and Corruption of the International Aid Business*. New York: Atlantic Monthly Press.

Hann, Chris, ed. 2002. *Postsocialism: Ideals, Ideologies, and Practices in Eurasia*. New York: Routledge.

Haraway, Donna. 1988. "Situated Knowledges: The Science Question in Feminism and the Privilege of Partial Perspective." *Feminist Studies* 14 (3): 575–99.

———. 1997. *Modest_Witness@Second_Millennium.FemaleMan_Meets_OncoMouse: Feminism and Technoscience*. New York: Routledge.

———. 2003. *The Companion Species Manifesto: Dogs, People, and Significant Otherness*. Paradigm 8. Chicago: Prickly Paradigm Press.

Harms, Erik. 2016. *Luxury and Rubble: Civility and Dispossession in the New Saigon*. Berkeley: University of California Press.

Harper, Richard. 1998. *Inside the IMF: An Ethnography of Documents, Technology and Organisational Action*. New York: Routledge.

Harvey, David. 2011. *A Brief History of Neoliberalism*. Reprint. Oxford: Oxford University Press.

Harvey, Penelope, Casper Bruun Jensen, and Atsuro Morita, eds. 2017. "Infrastructural Complications." In *Infrastructures and Social Complexity: A Companion*, 1–22. London: Routledge, Taylor & Francis Group.

Hayden, Cori. 2003. *When Nature Goes Public: The Making and Unmaking of Bioprospecting in Mexico*. Princeton, NJ: Princeton University Press.

Head, Chris. 2000. "Financing of Private Hydropower Projects." World Bank Discussion Papers. World Bank Group eLibrary. July 2000. http://elibrary.worldbank.org/doi/book/10.1596/0-8213-4799-3.

Heatherington, Kregg. 2011. *Guerrilla Auditors: The Politics of Transparency in Neoliberal Paraguay*. Durham, NC: Duke University Press.

Herzfeld, Michael. 1996. *Cultural Intimacy: Social Poetics in the Nation-State*. New York: Routledge.

Hickey, Samuel, and Giles Mohan, eds. 2004. *Participation, from Tyranny to Transformation?: Exploring New Approaches to Participation in Development*. New York: ZED Books.

High, Holly. 2008. "The Implications of Aspirations: Reconsidering Resettlement in Laos." *Critical Asian Studies* 40 (4): 531–50. doi:10.1080/14672710802505257.

———. 2014. *Fields of Desire: Poverty and Policy in Laos*. Singapore: National University of Singapore Press.

High, Holly, Ian G. Baird, Keith Barney, Peter Vandergeest, and Bruce Shoemaker. 2009. "Internal Resettlement in Laos: Reading Too Much into Aspirations: More Explorations of the Space between Coerced and Voluntary Resettlement in Laos." *Critical Asian Studies* 41 (4): 605–20. doi:10.1080/14672710903328039.

Hirsch, Philip. 2001. "Globalisation, Regionalisation and Local Voices: The Asian Development Bank and Rescaled Politics of Environment in the Mekong Region." *Singapore Journal of Tropical Geography* 22 (3): 237–51. doi:10.1111/1467-9493.00108.

Hirschman, Albert O. (1967) 1995. *Development Projects Observed*. Washington, DC: Brookings Institution.

Hochstetler, Kathryn, and Margaret E. Keck. 2007. *Greening Brazil: Environmental Activism in State and Society*. Durham, NC: Duke University Press.

Holbraad, Martin, Morten Axel Pedersen, and Eduardo Viveiros de Castro. 2014. "The Politics of Ontology: Anthropological Positions." Theorizing the Contemporary, Cultural Anthropology (website), January 13, 2014. https://culanth.org/fieldsights/462-the-politics-of-ontology-anthropological-positions.

Holmes, Douglas, and George E. Marcus. 2005. "Cultures of Expertise and the Management of Globalization: Toward the Re-functioning of Ethnography." In *Global Assemblages: Technology, Politics, and Ethics as Anthropological Problems*, edited by Aihwa Ong and Stephen J. Collier, 235–52. Oxford: Blackwell.

Howe, Cymene, and Dominic Boyer. 2016. "Aeolian Extractivism and Community Wind in Southern Mexico." *Public Culture* 28 (2): 215–35.

Huet, Marie-Hélène. 2013. *The Culture of Disaster*. Chicago: University of Chicago Press.

Hughes, David McDermott. 2006. "Hydrology of Hope: Farm Dams, Conservation, and Whiteness in Zimbabwe." *American Ethnologist* 33 (2): 269–87. doi:10.1525/ae.2006.33.2.269.

Hughes, Thomas. 1983. *Networks of Power: Electrification in Western Society, 1880–1930*. Baltimore, MD: John Hopkins University Press.

Humphreys, Caroline. 2002. "Does the Category 'Postsocialism' Still Make Sense?" In *Postsocialism: Ideals, Ideologies, and Practices in Eurasia*, 12–15. London: Routledge.

IAEA. 2001. *Present and Future Environmental Impact of the Chernobyl Accident*. Vienna: International Atomic Energy Agency. http://www-pub.iaea.org/MTCD/publications/PDF/te_1240_prn.pdf.

IRN. 1999. "Power Struggle: The Impacts of Hydro-Development in Laos." International Rivers (website). February 1, 1999. https://www.internationalrivers.org/resources/power-struggle-the-impacts-of-hydro-development-in-laos-4058.

International Rivers. 2010. "Existing and Planned Lao Hydropower Projects—Sept. 2010." International Rivers (website). http://www.internationalrivers.org/resources/existing-and-planned-lao-hydropower-projects-3527. March 26, 2010.

——. 2013. "Lancang River Dams: Threatening the Flow of the Lower Mekong." International Rivers. August 1, 2013. http://www.internationalrivers.org/files/attached-files/ir_lacang_dams_2013_5.pdf, accessed April 1, 2014.

Ivarsson S., T. Svensson, and S. Tønnesson. 1995. *The Quest for Balance in a Changing Laos: A Political Analysis*. NIAS Reports 25. Copenhagen: Nordic Institute for Asian Studies.

James, William. 2008. *Pragmatism: A New Name for Some Old Ways of Thinking*. Rockville, MD: Arc Manor.

Jensen, Casper Bruun, and Atsuro Morita. 2015. "Infrastructures as Ontological Experiments." *Engaging Science, Technology, and Society* 1 (November): 81–87. doi:10.17351/ests2015.007.

Jensen, Casper Bruun, and Brit Ross Winthereik. 2013. *Monitoring Movements in Development Aid: Recursive Partnerships and Infrastructures*. Infrastructures Series. Cambridge, MA: MIT Press.

Jerndal, Randi, and Jonathan Rigg. 1998. "Making Space in Laos: Constructing a National Identity in a 'Forgotten' Country." *Political Geography* 17 (7): 809–31. doi:10.1016/S0962-6298(98)00028-6.

Johns, Fleur E. 2015. "On Failing Forward: Neoliberal Legality in the Mekong River Basin." *Cornell International Law Journal* 48, 347–83.

Johnson, Andrew. 2013. "Progress and Its Ruins: Ghosts, Migrants, and the Uncanny in Thailand." *Cultural Anthropology* 28 (2): 299–319.

Keck, Frederic. 2008. "Food safety and animal diseases. The French Food Safety Agency, from mad cow disease to bird flu." *Medecine sciences: M/S* 24 (1): 81–86.

Keck, Margaret E., and Kathryn Sikkink. 1998. *Activists beyond Borders: Advocacy Networks in International Politics*. Ithaca, NY: Cornell University Press.

Kenney-Lazar, Miles. 2010. "Land Concessions, Land Tenure, and Livelihood Change: Plantation Development in Attapeu Province, Southern Laos." Vientiane: Faculty of Forestry, National University of Laos. https://mileskenneylazardotcom.files.wordpress.com/2015/11/kenneylazarlandconcessionsattapeu.pdf.

Kerr, Richard, and Richard Stone. 2009. "A Human Trigger for the Great Quake of Sichuan?" *Science* 323 (5912): 322. doi:10.1126/science.323.5912.322.

Kesby, Mike. 2005. "Retheorizing Empowerment-through-Participation as a Performance in Space: Beyond Tyranny to Transformation." *Signs: Journal of Women in Culture and Society* 30 (4): 2037–65. doi:10.1086/428422.

Khagram, Sanjeev. 2002. "Restructuring the Global Politics of Development: The Case of India's Narmada Valley Dams." In *Restructuring World Politics: Transnational Social Movements, Networks, and Norms*, edited by James V. Riker, Kathryn Sikkink, and Khagram, Sanjeev, 206–330. Social Movements, Protest, and Contention 14. Minneapolis: University of Minnesota Press.

——. 2004. *Dams and Development: Transnational Struggles for Water and Power*. Ithaca, NY: Cornell University Press.

——. 2005. "Beyond Temples and Tombs: Towards Effective Governance for Sustainable Development through the World Commission on Dams." In *International Commissions and the Power of Ideas*, edited by Ramesh Chandra Thakur, Andrew Fenton Cooper, and John English. New York: United Nations University Press.

Khagram, Sanjeev, James V. Riker, and Kathryn Sikkink, eds. 2002. *Restructuring World Politics: Transnational Social Movements, Networks, and Norms*. Social Movements, Protest, and Contention 14. Minneapolis: University of Minnesota Press.

Kingsnorth, Paul. 2015. *The Wake: A Novel*. Minneapolis, MN: Graywolf Press.

Kirksey, Eben. 2015. *Emergent Ecologies*. Durham, NC: Duke University Press.

Kirsch, Stuart. 2014. *Mining Capitalism: The Relationship between Corporations and Their Critics*. Berkeley: University of California Press.

——. 2010. "Sustainable Mining." *Dialectical Anthropology* 34 (1): 84–93. doi 10.1007/s10624-009-9113-x

Kleinman, Daniel Lee, and Sainath Suryanarayanan. 2013. "Dying Bees and the Social Production of Ignorance." *Science, Technology, & Human Values* 38 (4): 492–517. https://doi.org/10.1177/0162243912442575.

Kohn, Eduardo. 2013. *How Forests Think: Toward an Anthropology beyond the Human*. Berkeley: University of California Press.

Koopman, Colin. 2009. *Pragmatism as Transition: Historicity and Hope in James, Dewey, and Rorty*. New York: Columbia University Press.

Kothari, Uma. 2005. "Authority and expertise: The professionalisation of international development and the ordering of dissent." *Antipode* 37 (3): 425–46.

Kotter, John P. (1982) 1999. "What Effective General Managers Really Do." *Harvard Business Review*, March-April, 77 (2): 145–59.

Lahiri-Dutt, Kuntala. 2006. "Nadi O Nari: Social Construction of Rivers as Women in Rural Bengal." In *Fluid Bonds: Views on Gender and Water*, edited by Kuntala Lahiri-Dutt, 392–410. Kolkata, India: Stree.

Lahiri-Dutt, Kuntala, and Gopa Samanta. 2007. "'Like the Drifting Grains of Sand':1 Vulnerability, Security and Adjustment by Communities in the Char Lands of the Damodar River, India." *South Asia: Journal of South Asian Studies* 30 (2): 327–50. doi:10.1080/00856400701499268.

Langer, Paul. 1969. "Laos: Preparing for a Settlement in Vietnam." Rand Corporation (website). February 1969. http://www.rand.org/content/dam/rand/pubs/papers/2008/P4024.pdf.

Larkin, Brian. 2013. "The Politics and Poetics of Infrastructure." *Annual Review of Anthropology* 42: 327–43.

Latour, Bruno. 1987. *Science in Action: How to Follow Scientists and Engineers through Society*. Cambridge, MA: Harvard University Press.

——. 1993. *We Have Never Been Modern*. Cambridge, MA: Harvard University Press.

——. 1999. *Pandora's Hope: Essays on the Reality of Science Studies*. Cambridge, MA: Harvard University Press.

——. 2003. "Is Re-Modernization Occurring—And If So, How to Prove It?: A Commentary on Ulrich Beck." *Theory, Culture & Society* 20 (2): 35–48. doi:10.1177/0263276403020002002.

——. 2004. "Why Has Critique Run Out of Steam? From Matters of Fact to Matters of Concern." *Critical Inquiry* 30 (2): 225–48. doi:10.1086/421123.

——. 2013. *An Inquiry into Modes of Existence: An Anthropology of the Moderns*. Cambridge, MA: Harvard University Press.

Lave, Jean. 1988. *Cognition in Practice: Mind, Mathematics, and Culture in Everyday Life*. New York: Cambridge University Press.

Law, John. 1997. *The Manager and His Powers*. Lancaster, UK: Lancaster University, Department of Sociology.

Lee, Henry. 2006. "Financing the Theun-Hinboun Hydroelectric Project." Report #1829.0. Cambridge, MA: Harvard Kennedy School.

Li, Tania. 1999. "Compromising Power: Development, Culture, and Rule in Indonesia." *Cultural Anthropology* 14 (3): 295–322.

——. 2007. *The Will to Improve: Governmentality, Development, and the Practice of Politics*. Durham, NC: Duke University Press.

Lowe, Celia. 2006. *Wild Profusion: Biodiversity Conservation in an Indonesian Archipelago*. Information Series. Princeton, NJ: Princeton University Press.

Luhmann, Niklas. 1998. *Observations on Modernity*. Writing Science. Stanford, CA: Stanford University Press.

MacIntyre, Alasdair C. 1984. *After Virtue: A Study in Moral Theory*. 2nd ed. Notre Dame, IN: University of Notre Dame Press.
MacPhail, Theresa. 2010. "A Predictable Unpredictability: The 2009 H1N1 Pandemic and the Concept of 'Strategic Uncertainty' within Global Public Health." *Behemoth: A Journal of Civilization* 3 (3): 57–77.
Mallaby, Sebastian. 2006. *The World's Banker: A Story of Failed States, Financial Crises, and the Wealth and Poverty of Nations*, with a new afterword on the World Bank under Paul Wolfowitz. Council on Foreign Relations. New York: The Penguin Press.
Maltz, Maxwell. 1969. *Psycho-Cybernetics*. New York: Pocket Books.
Martin, Emily. 1994. *Flexible Bodies: Tracking Immunity in American Culture from the Days of Polio to the Age of AIDS*. Boston: Beacon Press.
Mbembe, Achille. 2001. *On the Postcolony*. Berkeley: University of California Press.
McLean, Stuart. 2009. "Stories and Cosmogonies: Imagining Creativity beyond 'Nature' and 'Culture.'" *Cultural Anthropology* 24 (2): 213–45. doi:10.1111/j.1548-1360.2009.01130.x.
——. 2011. "Black Goo: Forceful Encounters with Matter in Europe's Muddy Margins." *Cultural Anthropology* 26 (4): 589–619. doi:10.1111/j.1548-1360.2011.01113.x.
Merchant, Carolyn. 1980. *The Death of Nature: Women, Ecology, and the Scientific Revolution*. San Francisco: Harper & Row.
Mialet, Hélène. 2012. *Hawking Incorporated: Stephen Hawking and the Anthropology of the Knowing Subject*. Chicago: University of Chicago Press.
Michaud, Jean. 2013. "Comrades of Minority Policy in China Vietnam and Laos." In *Red Stamps and Gold Stars: Fieldwork Dilemmas in Upland Socialist Asia*, edited by Sarah Turner, 22ff. Vancover: UBC Press.
Milne, David. 2007. "'Our Equivalent of Guerrilla Warfare': Walt Rostow and the Bombing of North Vietnam, 1961–1968." *Journal of Military History* 71 (1): 169–203. doi:10.1353/jmh.2007.0056.
Mirzoeff, Nicholas. 2011. *The Right to Look: A Counterhistory of Visuality*. Durham, NC: Duke University Press.
Missingham, Bruce D. 2003. *The Assembly of the Poor in Thailand: From Local Struggles to National Protest Movement*. Chiang Mai, Thailand: Silkworm Books.
Mitchell Timothy. 1991. *Colonising Egypt*. Berkeley: University of California Press.
——. 2002. *Rule of Experts: Egypt, Techno-Politics, Modernity*. Berkeley: University of California Press.
Miyazaki, Hirokazu. 2013. *Arbitraging Japan: Dreams of Capitalism at the End of Finance*. Berkeley: University of California Press.
Mol, Annemarie. 2002. *The Body Multiple: Ontology in Medical Practice*. Durham, NC: Duke University Press.
Mongillo, John F., and Bibi Booth, eds. 2001. *Environmental Activists*. Westport, CT: Greenwood Press.
Moore, Aaron Stephen. 2013a. *Constructing East Asia: Technology, Ideology, and Empire in Japan's Wartime Era, 1931–1945*. Stanford, CA: Stanford University Press.
——. 2013b. "'The Yalu River Era of Developing Asia': Japanese Expertise, Colonial Power, and the Construction of Sup'ung Dam." *Journal of Asian Studies* 72 (1): 115–39. doi:10.1017/S0021911812001817.

———. 2014. "Japanese Development Consultancies and Postcolonial Power in Southeast Asia: The Case of Burma's Balu Chaung Hydropower Project." *East Asian Science, Technology and Society* 8 (3): 297–322. doi:10.1215/18752160-2416662.

Moore, Henrietta L. 1994. *A Passion for Difference: Essays in Anthropology and Gender*. Bloomington: Indiana University Press.

Moore, Sally Falk. 2001. "Certainties Undone: Fifty Turbulent Years of Legal Anthropology, 1949–1999." *Journal of the Royal Anthropological Institute* 7 (1): 95–116. doi:10.1111/1467-9655.00052.

Morita, Atsuro. 2016. "Infrastructuring Amphibious Space: The Interplay of Aquatic and Terrestrial Infrastructures in the Chao Phraya Delta in Thailand." *Science as Culture* 25 (1): 117–140.

Mosse, David. 2005. *Cultivating Development: An Ethnography of Aid Policy and Practice*. Anthropology, Culture, and Society. London: Pluto Press.

Murphy, Michelle. 2006. *Sick Building Syndrome and the Problem of Uncertainty: Environmental Politics, Technoscience, and Women Workers*. Durham, NC: Duke University Press.

———. 2008. "Chemical Regimes of Living." *Environmental History* 13 (4): 695–703.

———. 2013. "Chemical Infrastructures of the St Claire River." In *Toxicants, Health and Regulation since 1945*, 103–15. London: Pickering and Chatto.

Myers, Natasha. 2015. "Conversations on Plant Sensing: Notes from the Field." *NatureCulture* 3: 35–66.

Nelson, Paul. 2002. "New Agendas and New Patterns of International NGO Political Action." *Voluntas: International Journal of Voluntary and Nonprofit Organizations* 13 (4): 377–92.

Nguyen, Thi Dieu. 1999. *The Mekong River and the Struggle for Indochina: Water, War, and Peace*. Westport, CT: Praeger.

Obarrio, Juan. 2014. *The Spirit of the Laws in Mozambique*. Chicago: University of Chicago Press.

Ong, Aihwa. 1987. *Spirits of Resistance and Capitalist Discipline: Factory Women in Malaysia*. Albany: State University of New York Press.

———. 2000. "Graduated Sovereignty in South-East Asia." *Theory, Culture & Society* 17 (4): 55–75. doi:10.1177/02632760022051310.

———. 2006. *Neoliberalism as Exception: Mutations in Citizenship and Sovereignty*. Durham, NC: Duke University Press.

———. 2016. *Fungible Life: Experiment in the Asian City of Life*. Durham, NC: Duke University Press.

Oreskes, Naomi. 2014. *The Collapse of Western Civilization: A View from the Future*. New York: Columbia University Press.

Oreskes, Naomi, and Erik M. Conway. 2010. *Merchants of Doubt: How a Handful of Scientists Obscured the Truth on Issues from Tobacco Smoke to Global Warming*. New York: Bloomsbury Press.

Ottinger, Gwen. 2008. *Assessing Community Advisory Panels: A Case Study from Louisiana's Iindustrial Corridor*. Philadelphia: Chemical Heritage Foundation.

———. 2010. "Buckets of Resistance: Standards and the Effectiveness of Citizen Science." *Science, Technology & Human Values* 35 (2): 244–70. doi:10.1177/0162243909337121.

Pandian, Anand. Forthcoming. *A Possible Anthropology: Three Essays on Method*. Durham, NC: Duke University Press.

Pearson, Thomas. 2009. "On the Trail of Living Modified Organisms: Environmentalism within and against Neoliberal Order." *Cultural Anthropology* 24 (4): 712–45. doi:10.1111/j.1548-1360.2009.01045.x.

Petit, Pierre. 2002. "The Backstage of Ethnography as Ethnography of the State: Coping with officials in the Lao People's Democratic Republic." In *Red Stamps and Gold Stars: Fieldwork Dilemmas in Upland Socialist Asia*, edited by Sarah Turner, 143–64. Vancouver: UBC Press.

Petryna, Adriana. 2002. *Life Exposed: Biological Citizens after Chernobyl*. Princeton, NJ: Princeton University Press.

Pholsena, Vatthana. 2006. *Post-War Laos: The Politics of Culture, History, and Identity*. Ithaca, NY: Cornell University Press.

Pieterse, Jan Nederveen. 2000. "After Post-Development." *Third World Quarterly* 21 (2): 175–91. doi:10.1080/01436590050004300.

Pignarre, Philippe, and Isabelle Stengers. 2011. *Capitalist Sorcery: Breaking the Spell*. Houndmills, UK: Palgrave Macmillan.

Polletta, Francesca. 2004. *Freedom Is an Endless Meeting: Democracy in American Social Movements*. Paperback ed. Chicago: University of Chicago Press.

Povinelli, Elizabeth A. 2011. *Economies of Abandonment: Social Belonging and Endurance in Late Liberalism*. Durham, NC: Duke University Press.

——. 2016. *Geontologies: A Requiem to Late Liberalism*. Durham, NC: Duke University Press.

Rabinow, Paul. 1989. *French Modern: Norms and Forms of the Social Environment*. Cambridge, MA: MIT Press.

——. 1994. "Introduction: A Vital Rationalist." In *A Vital Rationalist: Selections of writings from Georges Canguilhem*, 11–23. Cambridge, MA: Zone Books.

——. 1996. *Essays on the Anthropology of Reason*. Princeton Studies in Culture/Power/History. Princeton, NJ: Princeton University Press.

——. 1997. "Introduction: The History of Systems of Thought." In *Ethics: Subjectivity and Truth*, edited by Paul Rabinow, xi–xlii. Vol. 1 of *Essential Works of Foucault 1954–1984*. London: Penguin.

——. 2003. *Anthropos Today: Reflections on Modern Equipment*. In-Formation Series. Princeton, NJ: Princeton University Press.

——. 2008. *Marking Time: On the Anthropology of the Contemporary*. Princeton, NJ: Princeton University Press.

Rabinow, Paul, and Gaymon Bennett. 2012. *Designing Human Practices: An Experiment with Synthetic Biology*. Chicago: University of Chicago Press. http://public.eblib.com/choice/publicfullrecord.aspx?p=914969.

Rabinow, Paul, and Nikolas Rose, eds. 2003. "Foucault Today." In *The Essential Foucault: Selections from Essential Works of Foucault, 1954–1984*, vii–xxxv. New York: New Press.

Raffles, Hugh. 2002. *In Amazonia: A Natural History*. Princeton, NJ: Princeton University Press.

——. 2017. "Against Purity." *Social Research: An International Quarterly* 84 (1): 171–82.

Rees, Tobias. 2018. *After Ethnos*. Durham, NC: Duke University Press.

Reynolds, Craig. 2005. "Power." In *Critical Terms for the Study of Buddhism*, edited by Donald S. Lopez, 211–28. Buddhism and Modernity. Chicago: University of Chicago Press.

Ribeiro, Gustavo. 1994. *Transnational Capitalism and Hydropolitics in Argentina: The Yacyretá High Dam*. Gainesville: University Press of Florida.

Rigg, Jonathan. 2009. "A Particular Place? Laos and Its Incorporation into the Development Mainstream." *Environment and Planning A* 41 (3): 703–21. doi:10.1068/a40260.

Riles, Annelise. 2000. *The Network Inside Out*. Ann Arbor: University of Michigan Press.

——. 2006. "[Deadlines]: Removing the brackets on politics in bureaucratic and anthropological analysis." In *Documents: Artifacts of Modern Knowledge*, edited by Annelise Riles, 71–92. Ann Arbor: University of Michigan Press.

Rose, Nikolas. 1999. *Powers of Freedom: Reframing Political Thought*. Cambridge: Cambridge University Press.

——. 2007. *The Politics of Life Itself: Biomedicine, Power, and Subjectivity in the Twenty-First Century*. Princeton, NJ: Princeton University Press.

Rostow, W. W. (1960) 1990. *The Stages of Economic Growth: A Non-Communist Manifesto*. Cambridge: Cambridge University Press.

Rottenburg, Richard. 2009. *Far-Fetched Facts: A Parable of Development Aid*. Inside Technology. Cambridge, MA: MIT Press.

Samimian-Darash, Limor. 2013. "Governing Future Potential Biothreats: Toward an Anthropology of Uncertainty." *Current Anthropology* 54 (1): 1–22. doi:10.1086/669114.

Samimian-Darash, Limor, and Paul Rabinow. 2015. "Introduction." In *Modes of Uncertainty: Anthropological Cases*, edited by Limor Samimian-Darash and Paul Rabinow, 1–12. Chicago: University of Chicago Press.

Sayre, Nathan. 2012. "The Politics of the Anthropogenic." *Annual Review of Anthropology* 41: 57–70

Schaper, Marcus. 2007. "Leveraging Green Power: Environmental Rules for Project Finance." *Business and Politics* 9 (3): 1–27. doi:10.2202/1469-3569.1184.

Scott, James C. 1990. *Domination and the Arts of Resistance: Hidden Transcripts*. New Haven, CT: Yale University Press.

——. 1998. *Seeing like a State: How Certain Schemes to Improve the Human Condition Have Failed*. Yale Agrarian Studies. New Haven, CT: Yale University Press.

Scranton, Roy. 2015. *Learning to Die in the Anthropocene: Reflections on the End of a Civilization*. San Francisco: City Lights Books.

Scudder, Thayer. 2006. *The Future of Large Dams: Dealing with Social, Environmental, Institutional and Political Costs*. Sterling, VA: Earthscan.

Shapin, Steven, and Simon Schaffer. 1985. *Leviathan and the Air-Pump: Hobbes, Boyle, and the Experimental Life*, including a translation of *Thomas Hobbes, Dialogus Physicus de Natura Aeris* by Simon Schaffer. Princeton, NJ: Princeton University Press.

Shoemaker, Bruce. 1998. *Trouble on the Theun Hinboun: A Field Report on the Socio-Economic and Environmental Effects of the Nam Theun-Hinboun Hydropower Project in Laos*. Berkeley, CA: International Rivers Network.

Shoemaker, Bruce, Ian G. Baird, and Monsiri Baird. 2001. *The People and Their River: A Survey of River-Based Livelihoods in the Xe Bang Fai River Basin in Central Lao PDR*. Vientiane, Laos: Canada Fund for Local Initiatives.

Singh, Sarinda. 2009. "World Bank-Directed Development? Negotiating Participation in the Nam Theun 2 Hydropower Project in Laos." *Development and Change* 40 (3): 487–507. doi:10.1111/j.1467-7660.2009.01562.x.

——. 2010. "Appetites and Aspirations: Consuming Wildlife in Laos: Appetites and Aspirations in Laos." *Australian Journal of Anthropology* 21 (3): 315–31. doi:10.1111/j.1757-6547.2010.00099.x.

——. 2011. "Bureaucratic Migrants and the Potential of Prosperity in Upland Laos." *Journal of Southeast Asian Studies* 42 (2): 211–31. doi:10.1017/S0022463411000026.

——. 2012. *Natural Potency and Political Power: Forests and State Authority in Contemporary Laos*. Honolulu: University of Hawai'i Press.

Sivaramakrishnan, K., and Arun Agrawal. 2003. "Regional Modernities in Stories and Practices of Development," In *Regional Modernities: The Cultural Politics of Development in India*, edited by K. Sivaramakrishnan and Arun Agrawal, 1–62. Stanford, CA: Stanford University Press.

Skafish, Peter. 2014. "Anthropological Metaphysics/Philosophical Resistance." Cultural Anthropology, Theorizing the Contemporary (website). Accessed January 13, 2014. https://culanth.org/fieldsights/464-anthropological-metaphysics-philosophical-resistance.

Smits, Mattijs, and Simon R. Bush. 2010. "A Light Left in the Dark: The Practice and Politics of Pico-Hydropower in the Lao PDR." *Energy Policy* 38 (1): 116–27. doi:10.1016/j.enpol.2009.08.058.

Sparke, Matthew. 2006. *Globalization and Paul Farmer's Reframing of Care*. Seattle: University of Washington Office of Minority Affairs and Diversity. https://geography.washington.edu/publications/globalization-and-paul-farmers-reframing-care.

Star, Susan Leigh, and Karen Ruhleder. 1996. "Steps toward an Ecology of Infrastructure: Design and Access for Large Information Spaces." *Information Systems Research* 7 (1): 111–34. doi:10.1287/isre.7.1.111.

Stengers, Isabelle. 2010. *Cosmopolitics I*. Minneapolis: University of Minnesota Press.

Stewart, Kathleen. 1996. *A Space on the Side of the Road: Cultural Poetics in an "Other" America*. Princeton, NJ: Princeton University Press.

Stonich, Susan C., and Peter Vandergeest. 2001. "Violence, Environment and Industrial Shrimp Farming." In *Violent Environments*, edited by Nancy Lee Peluso and Michael Watts, 261–86. Ithaca, NY: Cornell University Press.

Strathern, Marilyn. 1992. *After Nature: English Kinship in the Late Twentieth Century*. Cambridge: Cambridge University Press.

——. 2000. "The Tyranny of Transparency." *British Educational Research Journal* 26 (3): 309–21. doi:10.1080/713651562.

——. 2005. *Kinship, Law and the Unexpected: Relatives Are Always a Surprise*. New York: Cambridge University Press.

——. 2006. "Bulletproofing: A Tale from the United Kingdom." In *Documents: Artifacts of Modern Knowledge*, edited by Annelise Riles, 181–205. Ann Arbor: University of Michigan Press.

Stuart-Fox, Martin. 1993. "On the Writing of Lao History: Continuities and Discontinuities." *Journal of Southeast Asian Studies* 24: 106–21.

——. 1997. *A History of Laos*. Cambridge: Cambridge University Press.

———. 2005. "Politics and Reform in the Lao People's Democratic Republic." Working Paper No. 126. Asia Research Centre. Perth, Australia: Murdoch University.

Sunder Rajan, Kaushik. 2005. Subjects of Speculation: Emergent Life Sciences and Market Logics in the US and India. *American Anthropologist* 107 (1): 19–30.

Suyavong, Bounnakhone. 1997. "A Cost-Benefit Analysis of the Laos Nam Theun Two Project." PhD diss., University of Ottawa.

Swyngedouw, Erik. 2007. "Dispossessing H20: The Contested Terrain of Water Privatization." In *Neoliberal Environments: False Promises and Unnatural Consequences*, edited by Nik Heynen, James McCarthy, Scott Prudham, and Paul Robbins, 51–62. New York: Routledge.

Taillard, Christian. 1989. *Strategie de L'Etat-Tampon*. Paris: Reclus.

Tambiah, Stanley Jeyaraja. 1976. *World Conqueror and World Renouncer: A Study of Buddhism and Polity in Thailand against a Historical Background*. New York: Cambridge University Press.

Tanaka, Koji. 1993. "Farmers' Perceptions of Rice-Growing Techniques in Laos: 'Primitive' or 'Thammasat'?" *Southeast Asian Studies* 31 (3): 132–40.

Thongchai Winichakul. 1994. *Siam Mapped: A History of the Geo-Body of a Nation*. Honolulu: University of Hawai'i Press.

———. 2000. "The Quest for Siwilai: A Geographical Discourse of Civilizational Thinking in the Late Nineteenth and Early Twentieth-Century Siam." *Journal of Asian Studies* 59 (3): 528. doi: 10.1017/S0021911800014327.

Tooker, Deborah. 1996. "Putting the Mandala in Its Place: A Practice-Based Approach to the Spatialization of Power on the Southeast Asian 'Periphery'—The Case of the Akha." *Journal of Asian Studies* 55 (2): 323–58.

Tsing, Anna Lowenhaupt. 2005. *Friction: An Ethnography of Global Connection*. Princeton, NJ: Princeton University Press.

———. 2014. "Blasted Landscapes (and the Gentle Art of Mushroom Picking.)" In *The Multispecies Salon*, edited by Eben Kirksey, 87–109. Durham, NC: Duke University Press.

———. 2015. *The Mushroom at the End of the World: On the Possibility of Life in Capitalist Ruins*. Princeton, NJ: Princeton University Press.

Van den Berg, Rob D. 2004. "Evaluating the Fundamentalism of Evaluation." In *The Development of Religion, the Religion of Development*, edited by Ananta Kumar Giri, Anton van Harskamp, and Oscar Salemink, 67–73. Delft, the Netherlands: Eburon.

Vandergeest, Peter. 2003. "Land to Some Tillers: Development-Induced Displacement in Laos." *International Social Science Journal* 55 (175): 47–56. doi:10.1111/1468-2451.5501005.

Verdery, Katherine. 1996. *What Was Socialism, and What Comes Next?* Princeton, NJ: Princeton University Press.

Verran, Helen. 2012. "Number." In *Inventive Methods: The Happening of the Social,* edited by Celia Lury and Nina Wakeford, 110–24. New York: Routledge.

———. 2014. "Anthropology as Ontology is Comparison as Ontology." Theorizing the Contemporary. Cultural Anthropology (website). Accessed January 13, 2014. https://culanth.org/fieldsights/468-anthropology-as-ontology-is-comparison-as-ontology.

Viktorin, Mattias. 2008. *Exercising Peace: Conflict Preventionism, Neoliberalism, and the New Military*. Stockholm Studies in Social Anthropology 63. Stockholm: Department of Social Anthropology, Stockholm University.

Viveiros de Castro, Eduardo. 2014. *Cannibal Metaphysics: For a Post-Structural Anthropology*. Translated by Peter Skafish. Minneapolis, MN: Univocal.

Vorapeth, Kham. 2007. *Laos: Le Redefinition Des Strategies Politiques et Economiques (1975–2006)*. Paris: Indes Savantes.

Weheliye, Alexander. 2014. *Habeas Viscus: Racializing Assemblages, Biopolitics, and Black Feminist Theories of the Human*. Durham, NC: Duke University Press.

Weiss, Jessica. 2016. "Miami Beach's $400 Million Sea-Level Rise Plan Is Unprecedented, But Not Everyone Is Sold." Miami New Times, April 19. http://www.miaminewtimes.com/news/miami-beachs-400-million-sea-level-rise-plan-is-unprecedented-but-not-everyone-is-sold-8398989.

Welker, Marina A. 2009. "'Corporate Security Begins in the Community': Mining, the Corporate Social Responsibility Industry, and Environmental Advocacy in Indonesia." *Cultural Anthropology* 24 (1): 142–79. doi:10.1111/j.1548-1360.2009.00029.x.

West, Paige. 2006. *Conservation Is Our Government Now: The Politics of Ecology in Papua New Guinea*. Durham, NC: Duke University Press.

White, Gilbert F. 1963. "The Mekong River Plan." *Scientific American*, April 1, 208 (4), 310–16.

Whitington, Jerome. 2012. "The Institutional Condition of Contested Hydropower: The Theun Hinboun–International Rivers Collaboration." *Forum for Development Studies* 39 (2): 231–56. doi:10.1080/08039410.2012.666762.

——. 2013. "Fingerprint, Bellwether, Model Event: Climate Change as Speculative Anthropology." *Anthropological Theory* 13(4): 308–28.

——. 2014. "Laos." In *Figures of Southeast Asian Modernity*, edited by Joshua Barker, Erik Harms, and Johan Lindquist, 91–94. Honolulu: University of Hawai'i Press.

Williams, Philip. 1978. "Dam Design—Is the Technology Faulty?" *New Scientist* 77 (1088): 280–82.

Williams, Philip B. 1997. "The Experience of International Rivers Network, 1985–1997." Paper presented to the First International Conference of Peoples Affected by Dams. Couritiba: Brazil, March 10.

Winner, Langdon. 1980. "Do Artifacts Have Politics?" *Daedalus* 109 (1): 121–36.

Wolters, O. W. 1999. *History, Culture, and Region in Southeast Asian Perspectives*. Rev. ed. Studies on Southeast Asia 26. Ithaca, NY: Southeast Asia Program Publications, Southeast Asia Program, Cornell University.

Wong, Soo Mun Theresa. 2010. "Making the Mekong: Nature, Region, Postcoloniality." PhD diss., Ohio State University.

Woodrow, Robert. 1998. "Laos in 1997." Encyclopaedia Britannica (website). Accessed March 12, 2015. https://www.britannica.com/place/Laos-Year-In-Review-1997.

Woolgar, Steve. 2005. "Ontological Disobedience?–absolutely![perhaps]." In *The Disobedient Generation: Social Theorists in the Sixties*, edited by Alan Sica and Stephen P. Turner, 309–24. 1st ed. Chicago: University of Chicago Press.

Index